T0135636

AURALISATION IN BUILDING ACOUSTICS

Von der Fakulät für Elektrotechnik und Informationstechnik der
Rheinisch-Westfälischen Technischen Hochschule Aachen
zur Erlangung des akademischen Grades eines
DOKTORS DER INGENIEURWISSENSCHAFTEN
genehmigte Dissertation

vorgelegt von

Diplom-Ingenieur
Rainer Thaden
aus Aurich

Berichter: Universitätsprofessor Dr. rer. nat. Michael Vorländer
Universitätsprofessor Dr.-Ing. Peter Vary

Tag der mündlichen Prüfung: 09. Februar 2005

Bibliografische Information Der Deutschen Bibliothek

Die Deutsche Bibliothek verzeichnet diese Publikation in der Deutschen Nationalbibliografie; detaillierte bibliografische Daten sind im Internet über http://dnb.ddb.de abrufbar.

ISBN 3-8325-0895-3

Logos Verlag Berlin
Comeniushof, Gubener Str. 47,
10243 Berlin
Tel.: +49 030 42 85 10 90
Fax: +49 030 42 85 10 92
INTERNET: http://www.logos-verlag.de

Contents

Abstract – Zusammenfassung

Abstract

In this thesis, an algorithm for auralisation of airborne and impact sound insulation in buildings was developed, verified, and applied in several areas of research. After a short introduction to prediction methods for sound insulation in European countries, two methods for calculation of sound insulation in a two-room situation are described. These methods use the data of the building products to calculate the total sound insulation. The results form the input data for the auralisation algorithm. After the description of the basic principles of auralisation in general and fundamentals of the algorithms for auralisation of airborne and impact sound insulation, a verification regarding psychoacoustic and subjective parameters is presented.

The practical relevance of building acoustical auralisation is proven with two applications of the auralisation of airborne sound insulation. Both applications deal with speech transmitted through sound insulating constructions. First, speech intelligibility was investigated. It plays an important role for acoustic comfort (privacy in dwellings) and confidentiality in office buildings. In a second application, the influence on mental processing was evaluated.

Building acoustical auralisation is useful in many fields of applications. It can be used in the design of buildings or in teaching. Students of acoustics, architecture, and civil engineering get an impression of the acoustic effects of different materials and construction methods immediately. In research, it allows the execution of listening tests without having to carry out extensive measurements and recordings. Thus, it is a useful tool to improve the quality of reference curves for single number ratings.

Zusammenfassung

Im Rahmen der Arbeit wurde ein Algorithmus zur Auralisation (Hörbarmachung) der Schalldämmung in Gebäuden entwickelt, verifiziert und in unterschiedlichen Anwendungsgebieten erprobt. Nach einer kurzen Übersicht über die Vorhersagemethoden der Schalldämmung in den europäischen Ländern folgt die Beschreibung zweier Verfahren zur Berechnung der Schalldämmung einer 2-Raum-Situation anhand der Daten der verwendeten Bauprodukte. Diese Verfahren liefern die Eingangsdaten des verwendeten Auralisationsalgorithmus. Nach der ausführlichen Darlegung der Grundlagen und der Algorithmen zur Auralisation der Luft- und Trittschalldämmung wird eine Verifikation anhand psychoakustischer Auswertungen und Hörversuchen beschrieben.

Die praktische Relevanz des Verfahrens wird anhand zweier durchgeführter Anwendungen des Algorithmus zur Luftschall-Auralisation verdeutlicht. Die erste Anwendung befasst sich mit der Verständlichkeit von Sprache, welche durch schalldämmende Konstruktionen übertragen wird. Der Bezug zur Praxis liegt hier z.B. beim akustischen Komfort (Privatsphäre in Wohnungen) und der Wahrung der Vertraulichkeit in Büroräumen. Des weiteren wurde die Beeinflussung der mentalen Verarbeitung visueller Inhalte (Arbeitsgedächtnis) durch sprachlichen Hintergrundschall untersucht.

Das Verfahren der bauakustischen Auralisation lässt sich in vielerlei Hinsicht einsetzen. Es erleichtert die bauakustische Planung und kann in der Lehre eingesetzt werden, um Studenten der Akustik, der Architektur und des Bauingenieurwesens die Wirkung des Einsatzes bestimmter Baustoffe und schalltechnischer Maßnahmen unmittelbar vorzuführen. In der Forschung erlaubt die Auralisation, umfangreiche Hörversuche durchzuführen und somit verbesserte Bewertungskurven für die Angabe von Schallwirkungen in Gebäuden zu ermitteln.

Keywords

building acoustics, auralisation, sound insulation, impact sound

Chapter 1

Introduction

In modern societies, especially in urban areas, the exposure of man to noise is steadily growing. This, of course, is related to aspects as mobility and technical progress which are responsible for growth of traffic and increased use of electrical and mechanical devices such as air conditions, ventilations, elevators, hifi systems and home theaters. Road freight transport, for example, took an increase of 41% between 1991 and 2000. A survey authorised from the Federal Environmental Agency of Germany in 2000 states that about 18% of the german population is highly annoyed by traffic noise [OW02]. An important point of concern is the high annoyance of 6.5% of the population by noise of neighbours. Neighbourhood noise is amongst traffic, airplane, railway, and industry noise one of the main causes for annoyance. In total, only 20% of the people state not to be annoyed by noise. Taking a detailed look into the survey, one notes that particularly people living in multi family and apartment houses are annoyed by noise of their neighbours. This leads to the conclusion that mainly noise caused by indoor activities is responsible for the annoyance. Also, it can be assumed that insufficient requirements in the german standard DIN 4109 *Sound insulation in buildings* play an important role since more than 80% of the buildings are built according to the standard. A literature study by Rasmussen and Rindel [RR96] of surveys in different countries about noise protection, noise from neighbours, and needs of habitants states that

- many people are annoyed by noise from neighbours.

- people are annoyed by the fact that their neighbours can hear them.
- habitants are prepared to pay about 2-3% more rent to improve noise protection.
- the degree of annoyance is influenced by the personal conditions of the habitants (stress), the value of the dwelling and the duration the habitants live there.
- people judge their neighbours as more or less considerate according to the quality of sound insulation. This may cause social conflicts in buildings with poor sound insulation.
- airborne sound insulation should exceed R'_w = 60 dB and the impact noise level should be below L'_n = 48 dB to be judged as satisfactory by the inhabitants and keep the number of complaints below a reasonable level.

There are different consequences arising out of noise exposure [Gus02]:

- Masking of the speech signal by noise causes a disturbance of communication. There is an increased effort needed for listening and talking. This may disturb conversation or listening to the TV or radio in private dwellings as well as communication in office premises.
- Reduced power of concentration: Experiments with adults have shown that physical or mental work is impaired by noise, in particular if high concentration is required. Very demanding activities can be disturbed by levels of already 45 dB(A). In various studies, pupils showed an impairment of reading and memorising items of speech when exposed to aircraft noise. Also, exposure to road traffic and railway noise resulted in these effects.
- Disturbance of sleep: Sleep consists of different phases which effect physical and mental regeneration. Experiments where an EEG and signals of the respiration, circulation and body movements are recorded during sleep show that noise exposure affects the sequence and duration of these phases and can lead to decreased cognitive and physical performance during the day.

Thus, it can be seen that reduction of noise exposure is not only a factor for the quality of life regarding acoustic comfort in dwellings but also a socio-economical cost factor since noise exposure

- can reduce productivity and cause human failure
- affects human health leading to increased costs of the health system
- may cause loss of property value if, for example, aircraft noise increases due to a new runway in the area.

How to assess the influences of noise exposure and how to reduce its effects with not only acoustical means is subject of research on noise effects in psycho and socio acoustics.

It is the responsibility of the government to protect citizens from noise exposure by means of legal regulations which limit noise levels and set standards for quantities as e.g. airborne and impact sound insulation. These regulations must not only define noise limits not to be exceeded but must, for example, also define standardised data for description of building products and measurement procedures to assess the data. This allows manufacturers to distribute their building products and eases work for architects and civil engineers. The performance of a newly constructed building has to be predicted before construction is begun. Since the requirements for minimum sound insulation often specify measurable quantities in the already constructed building, the prediction methods have to be based on data of the used materials and products.

In 1988, the Construction Products Directive has been passed by the European Commission. In the interpretative document 'Protection against noise', several needs for standardisations are specified. The technical committee CEN/TC 126 (Comité Européen de Normalisation) was installed to specify the necessary requirements in several working groups. A list of the working groups is shown in Table 1.1. The objective was to provide directives and regulations to ease trading in the European Union. The work of WG 2 resulted in the European standard EN 12354 'Prediction of the acoustic performance of buildings from the performance of products' which will be discussed in detail in chapter 2.3. The standard describes the calculation of sound insulation of buildings from their building products and

WG 1	Methods for measuring the sound insulation of building elements and the performance of buildings
WG 2	Prediction of the acoustic performance of buildings from the performance of products
WG 3	Laboratory test on noise from hydraulic equipment used in water installation
WG 4	Single number ratings of the acoustic performance of buildings and building products
WG 5	Coordination
WG 6	Laboratory measurements of flanking transmission
WG 7	Laboratory measurements of noise from waste water installations
WG 8	Field measurement of reverberation time

Table 1.1: Working groups in CEN/TC 126

defines the data necessary for calculation. Thus, it is possible for the building acoustic consultant to use a software with a database of building products and their relevant data to virtually construct buildings and have the resulting sound insulation calculated by the implemented prediction model. This, of course, is an improvement over the method used in DIN 4109 [DIN89] which is still in use at present. DIN 4109 supplies a list of constructions with estimations of their performance and rules of thumb for estimating the sound insulation. A short description of prediction models used in different European countries can be found in chapter 2.1.

The resulting quantities from the newer prediction models or measurements are normally frequency dependent. To allow a more practicable use, methods are required which transform the frequency dependent quantities to single numbers. In Germany, in the 1950s, the mean value of the frequency dependent quantity was used. This was revealed as unsatisfactory because, for example, poor insulation at low frequencies could be compensated with good insulation at high frequencies. Since the early 1950s, a method proposed by Cremer using reference curves is used in Germany and became an ISO standard [ISO97] in 1968. Another method, used e.g. in France and Scandinavia is the single number rating based on A-weighted level differences using defined spectra of indoor and outdoor noise. Since the use of different reference curves leads to different ratings, it has to be evaluated carefully which curve is to be standardised. Lang [Lan97] reports that in listening tests during work on standardisation in CEN/TC 126 WG4 it was found that sound transmitted over two windows with equal weighted sound reduction indexes R_w but different weighted sound reduction indexes with spectrum adaptation terms $R_{A,tr}$ was perceived with different loudness whereas sound transmitted over two windows with equal $R_{A,tr}$ but different R_w was considered to have the same loudness. To consider also the method proposed in France and Scandinavia, it was decided to supply two spectrum adaptation terms C and C_{tr} which measure the difference $R_A - R_w$ resp. $R_w - R_{A,tr}$. It can be seen that listening tests provide useful information for the finding of objective criteria.

For testing of impact sound insulation, the use of the standard tapping machine is required according to ISO 140-7. For single-number rating, there exists a method similar to the one for airborne sound insulation which is described in [ISO97]. The use of the tapping machine has some serious drawbacks:

- There is a feedback from the structure to the exciting source making comparison between heavy-weight and light-weight constructions problematic as well as the comparison of floor coverings on different constructions.
- The sound level spectrum in the receiving room caused by the tapping machine does not match with the spectrum caused by the most common noise sources like walking persons or jumping and running children.

Thus, the execution of listening tests is difficult as the signal of the tapping machine is not usable. Other sources have to be used which should be standardised to allow a comparison between different studies, e.g., the rubber ball or the so called 'bang machine' which were developed in Japan. There is a strong discussion going on at the time being about the correct measurement method and the used source. A modification to the tapping machine has been proposed by Scholl [SWSK02] to deliver comparable results for light-weight and heavy-weight structures. Investigations by Fischer [FDS03] and Petzold [Pet02] showed that this modification yields sound level spectra comparable to those of human walkers. Listening tests on psycho-acoustic parameters of impact sound were carried out, e.g., by Jeon et al. [JJVT04] and Johansson et al. [JHN04].

It has to be investigated if test procedures and rating methods for airborne and impact sound insulation can be found which better consider the subjective impressions of subjects and psychoacoustic and psychological factors which can not be measured physically. To obtain data for this investigation, listening tests have to be carried out with sound signals in the receiving room. Normally, the audio signals necessary are recorded in real situations which is very time consuming and expensive. It would be desirable to have a simulation tool which calculates the sound field at the ears of the listener from predicted or measured data. This can be achieved by a method called auralisation.

The method of auralisation is known since the early 20th century when Spandöck [Spa34] made first experiments with scale models. Auralisation means to make a calculated or measured sound field audible. In room acoustics, it is used to get an impression of the acoustic behaviour of a room, e.g. a concert hall in an early stage of development or before construction works. In general, an impulse response describing the propagation of sound from a source to a receiver is determined by measurement or calculation. Then, a convolution of a sound source with this impulse response delivers the sound signal at the receiver. In room acoustics, normally, room impulse responses for a certain source receiver configuration are calculated by room acoustic simulation software using ray tracing, for example. The room impulse response can also be obtained by a measurement in an already existing room. The auralisation, then, only consists of the convolution of the source signal with the impulse response. This way it is possible, e.g. to place a singer or an orchestra into different concert halls without having to travel. Dry source signals are needed for this method, i.e. signals from sources recorded in an anechoic room and also impulse responses of the rooms to be auralised.

Whereas in room acoustics the modeling of the path consists only of considering the sound propagation in air, the modelling of the source receiver path needs to consider the propagation of sound through structures. For a simple room-to-room situation, it is necessary to model the propagation of the sound in both rooms, the excitation of and radiation from the walls and the properties of the receiver. To reduce the complexity of the model to a reasonable amount, some simplifications are necessary. Work on building acoustical auralisation has been done, for example, by Schmitz and Zier [SZ97] and Lehn et al [LSB98]. A general overview of methods and applications of auralisation in noise control was given by Vorländer [Vor03].

To get an impression of the importance and usefulness of auralisation, one can refer to visualisation. Pictures can be expressed by characteristics like hue, saturation, brightness, and maybe by descriptions of the objects and their arrangement. They can, then, be compared according to those values. Now, imagine to explain the Mona Lisa of Leonardo da Vinci (see Figure 1.1) to somebody who has never seen the painting. And further, try to explain the differences between pictures of Leonardo da Vinci and Vincent van Gogh. Of course, references to other paintings, artists, or styles may be helpful but only if the person has knowledge of painting. If single numbers are used, lots of them are needed and they will not be too helpful. How easy would it be to give people an impression by just showing the picture to them.

Figure 1.1: How to express differences between these two pictures with single-number ratings such as mean colour, mean brightness, hue, saturation etc.?

In this thesis, algorithms for auralisation of airborne and impact sound insulation have been developed and implemented. Chapter 2 gives an overview of models for predicting sound insulation which deliver the input data for building acoustic auralisation. In particular, the model from EN 12354 is discussed for it was used in this work. Chapter 3 describes the principles of auralisation including signal processing and binaural techniques. Chapters 4 and 5 describe in detail the models used for auralisation of airborne and impact sound insulation. Models for sound propagation in both rooms and through structures are explained and the simplifications which were made to reduce the complexity of the model but still deliver an authentic sound field. It has to be verified if the sound field is authentic. This has been done not only for the correct levels which are reproduced at the ears of the listener but also for subjective responses of subjects. Therefore, sound signals were recorded at real room situations as well as auralised and compared regarding acoustic parameters and subjective impressions. The methods and results are described in chapter 6. Also, some applications of auralisation in building acoustics have been tested. Descriptions of these applications as rating of speech intelligibility of different room situations or evaluating disturbance effects on concentration in large offices with separating walls (Irrelevant Speech Effect) can be found in chapter 7. A summary of this thesis and an outlook on future work is given in chapter 8.

Chapter 2

Prediction models of building acoustics

To predict the performance of sound insulation of a room situation from the data of its building products, models are necessary which are fed with quantities describing the construction and the materials. There exist several models, some of which are rather specialized to certain situations whereas others are kept as general as possible to be applicable for a wide range of building methods and constructions. This chapter introduces the quantities used for characterisation of airborne and impact sound insulation in two subsections and finally gives an overview over common modern models as the statistical energy analysis and the method used in EN 12354 [EN197] as well as traditional models as used in some European countries.

2.1 Prediction Models used in European countries

There exist different prediction models in the European countries which, however, mostly use the same input quantities for the description of building products but different quantities for rating the sound insulation of buildings. The material, normally, is described by the sound reduction index

$$R = L_S - L_R + 10 \log \frac{S}{A} \tag{2.1}$$

with L_S and L_R denoting the levels measured in a testing facility in the source and receiving room, S denoting the area of the partition, and A the equivalent absorption area of the receiving room. R is originally derived from the relation of the sound intensities in the free source and receiving domain but can be expressed by sound pressure levels as above with the assumption of diffuse fields in both rooms and the consideration of S and A. If R, however, is measured in buildings, then the sound transmitted over the flanking walls (flanking transmission) is included in the measurement. Thus, it can not be considered as a description of a single building product. The resulting quantity, then, is called apparent sound reduction index R'.

A second approach is the description by the level difference in both rooms without a relation to the area of the partition. This, actually, describes the sound insulation of the whole room situation. The normalised level difference

$$D_n = L_S - L_R - 10 \log \frac{A}{10\ \mathrm{m}^2} \tag{2.2}$$

and the standardised level difference

$$D_{nT} = L_S - L_R + 10 \log \frac{T}{0.5\ \mathrm{s}} \tag{2.3}$$

differ only in the reference to the sound field in the receiving room. D_n refers the equivalent absorption area to 10 m^2 and D_{nT} refers the reverberation time to 0.5 s, both being typical values for average living rooms.

In practice, single numbers derived from these frequency-dependent quantities are used. Basically, there are two different approaches for single number rating of buildings and products:

- comparison with a reference curve: In most countries the procedure from ISO 717 [ISO97] is used to obtain R_w, R'_w, $D_{n,w}$, $D_{nT,w}$, $L_{n,w}$, or $L_{nT,w}$.

- A-weighted level difference R_A/D_{nAT} or L_{nAT}: A reference spectrum in the source room is used and the difference of the A-weighted levels in the source and receiving room is calculated. Reference spectra exist for indoor noise (used for C) and inner city traffic noise (used for C_{tr}) in octave or one-third octave bands. The receiving room level is calculated from the sound insulation quantity (R, D_{nT} or L_{nT}) and the reference spectrum.

$$R_A = -10\log\left(\sum_{i=1}^{n} 10^{(L_i - R_i)/10}\right)$$
$$R_{A,tr} = -10\log\left(\sum_{i=1}^{n} 10^{(L_{i,tr} - R_i)/10}\right) \quad (2.4)$$

where L_i, $L_{i,tr}$ are the values of the reference spectra. To reference this quantity to the weighted quantities from ISO717, spectrum adaptation terms C and C_{tr} have been introduced in a way that

$$C = R_A - R_w$$
$$C_{tr} = R_{A,tr} - R_w \quad (2.5)$$

In the following the prediction model used up to the time in Germany shall be described in short to give the reader an impression on the basic principle of 'older methods' using single numbers, correction terms, and rules of thumb. Most of the following content is taken from an article by Metzen [Met92]. For a more detailed overview and references to the standards of the different countries, the reader is referred to that article.

Germany (DIN4109) In the German standard DIN 4109 [DIN89], the demanded minimum sound insulation between dwellings is given by values of R'_w = 53 dB resp. 54 dB (walls/floors) and the maximum impact sound level is 53 dB. To predict these values from construction data, tables from DIN4109 Appendix 1 can be used which specify values of R'_w for different constructions:

- single homogeneous walls with/without linings
- double homogeneous walls
- double light-weight partitions with framework
- massive floors with/without floating floors
- wooden floors

It is assumed that the flanking walls have a mean mass per area of 300±25kg. For situations with different flanking conditions, a correction term $K_{L,1}$ is used. Another correction term $K_{L,2}$ is used for situations with different linings or floating floors at flanking paths. As an example: For a wall with a mass per area of 400 kg an $R'_{w(300)}$ = 53 dB can be read from a diagram. If the mean mass per area of the flanking paths is 200 kg the correction term $K_{L,1}$ can be determined to −1 dB. The sound insulation can, thus, be determined to $R'_w = R'_{w(300)} + K_{L,1} + K_{L,2} = 53\text{dB} - 1\text{dB} + 0\text{dB} = 52\text{dB}$.

The determination of the impact sound insulation is stuck to constructions with floating floors. The weighted impact sound level $L'_{n,w}$ results from the difference between the impact sound level of the floor and the floor covering. This yields $L'_{n,w} = L'_{n,w,eq} - \Delta L_w$ with $L'_{n,w,eq}$ as the equivalent weighted normalised impact sound level and ΔL_w as the impact sound level improvement of the floor covering. There is no correction for different flanking path conditions since they are considered as not having much impact on floating floor constructions.

The method used in DIN 4109 is mainly based on relations obtained from empirical data (measurements in buildings). The advantage of this method is its usability by a broad range of users. On the other hand, the method is stuck to only few traditional building styles which existed at the time of its creation. More modern styles are hard to handle.

It is criticised that the requirements in Germany are too low since many people complain about unsatisfactory noise protection. This may be due to the trust of people in a good noise protection guaranteed by the standard. The standard, however, only defines levels ‘to protect occupied areas against offending noise’. It can not be expected that noise from outdoors or adjoining rooms are undetectable [Poh03]. In a revised version of the standard DIN 4109-10(E):2002-07, the relevant quantities for sound insulation are changed from $R'_w/L'_{n,w}$ to $D_{nT,w}/L'_{nT,w}$. Also, a concept with three classes of acoustical comfort is taken over from the directive VDI4100. Since the calculation in the old DIN 4109 is still based on data from test facilities with a typical flanking transmission which, however, is not in accordance with the new European concept, a new building products catalogue with compatible data is being created at the time of writing. Another point of criticism is the

Country and quantity for requirement formulation		**Multi-storey**		**Terraced**	
		Req. [dB]	Eq. R'_w	Req. [dB]	Eq. R'_w
Austria	$D_{nT,w}$	55	~54-57	60	~59-62
Belgium	$D_{nT,w}$	54	~53-56	58	~57-60
Czech Rep.	R'_w	52	52	57	57
Denmark	R'_w	52	52	55	55
Estonia	R'_w	55	55	55	55
Finland	R'_w	55	55	55	55
France	$D_{nT,w} + C$	53	~53-56	53	~53-56
Germany	R'_w	53	53	57	57
Hungary	R'_w	52	52	57	57
Iceland	R'_w	52	52	55	~55
Italy	R'_w	50	50	50	50
Latvia	R'_w	54	54	54	54
Lithuania	$D_{nT,w}$ or R'_w	55	~55	55	~55
Netherlands	$I_{lu;k}$	0	~55	0	~55
Norway	R'_w	55	55	55	55
Poland	$R'_w + C$	50	~51	52	~53
Portugal	$D_{n,w}$	50	~50-52	50	~50-52
Russia	I_b	50	52	No requirements	
Slovakia	R'_w	52	52	52	52
Slovenia	R'_w	52	52	52	52
Spain	$D_{nT,w} + C_{100-5000}$	50	~50-53	50	~50-53
Sweden	$R'_w + C_{50-3150}$	53	~55	53	~55
Switzerland	$D_{nT,w} + C$	54	~54-57	54	~54-57
UK	$D_{nT,w} + C_{tr}$	45	~49-52	45	~49-52

Table 2.1: Airborne sound insulation requirements of 24 European countries. From [Ras04]

omission of low frequencies, especially for impact sound. If floating floors are used, the resonance frequencies are often below the lowest considered frequency of 100 Hz. The resulting booming noise causes annoyance which the dwellers have no means against [SF03].

European countries Tables 2.1 and 2.2 give an overview of quantities commonly used for characterising airborne and impact sound insulation in 24 European countries. In some countries R'_w is used for the description of sound insulation between dwellings and in some countries $D_{nT,w}$ is used. Since $D_{nT,w}$ is describing the level difference and not the separating

Country and quantity for requirement formulation		Multi-storey Req. [dB]	Multi-storey Eq. $L'_{n,w}$	Terraced Req. [dB]	Terraced Eq. $L'_{n,w}$
Austria	$L'_{nT,w}$	48	~50-43	46	~48-41
Belgium	$L'_{nT,w}$	58	~60-53	50	~52-45
Czech Rep.	$L'_{n,w}$	58	58	53	53
Denmark	$L'_{n,w}$	58	58	53	53
Estonia	$L'_{n,w}$	53	53	53	53
Finland	$L'_{n,w}$	53	53	53	53
France	$L'_{nT,w}$	58	~60-53	58	~60-53
Germany	$L'_{n,w}$	53	53	48	48
Hungary	$L'_{n,w}$	55	55	47	47
Iceland	$L'_{n,w}$	58	58	53	53
Italy	$L'_{n,w}$	63	63	63	63
Latvia	$L'_{n,w}$	54	54	54	54
Lithuania	$L'_{n,w}$	53	53	53	53
Netherlands	I_{co}	+5	~61-54	+5	~61-54
Norway	$L'_{n,w}$	53	53	53	53
Poland	$L'_{n,w}$	58	58	53	53
Portugal	$L'_{n,w}$	60	60	60	60
Russia	I_y	67	60	No requirements	
Slovakia	$L'_{n,w}$	58	58	58	58
Slovenia	$L'_{n,w}$	58	58	58	58
Spain	$L'_{nT,w}$	65	~67-60	65	~67-60
Sweden	$L'_{n,w} + C_{i,50-2500}$	56	~56	56	~56
Switzerland	$L'_{nT,w} + C_i$	50	~52-45	50	~52-45
UK	$L'_{nT,w}$	62	~64-57	None	Not avail.

Table 2.2: Impact sound requirements of 24 European countries. From [Ras04]

construction element and the area of this element as well as the flanking walls are taken into account, it seems to be more appropriate.

As can be seen from the descriptions above, the prediction models in the European countries are somewhat different in both, the requirements to be met by the constructions and the calculation method. Also, the quantity used for rating the sound insulation of the constructions is different. Table 2.3 summarises the number of different concepts and quantities showing how far the directives are away from a harmonised concept. It can, however, be stated that at least all countries can use the same input quantities as described

Airborne sound insulation	Impact sound insulation
10 concepts + variants / recommendations	6 concepts + variants / recommendations
Multi-storey housing: Variation of 6 dB in eq. R'_w	Multi-storey housing: Variation of 17 dB in eq. $L'_{n,w}$
Terraced housing: Variation of 11 dB in eq. R'_w	Terraced housing: Variation of 19 dB in eq. $L'_{n,w}$
Stricest requirement in Austria	Strictest requirement in Austria

Table 2.3: Comparison of requirements in 24 European countries. From [Ras04]

by EN 12354 for the calculation of the relevant quantities for formulating the requirements. Whereas most of the countries have their calculations based on single numbers, the countries with newly developed standards took a step further to using frequency-dependent data. Frequencies below 100 Hz are dealt with in different ways. In the UK, e.g., the low frequencies are covered by using the spectrum adaptation terms C and C_{tr} but without using input data for sound insulation below 100 Hz. This seems inappropriate, since a correction term is used on data which is not present at all.

The use of building product catalogues (as, e.g., used in Germany) is a very static method since newly developed building methods have to be included before use and, e.g., different flanking conditions may not be covered. Also frequency-dependent data should be used for calculations instead of single number ratings. It is desirable to have prediction models which cover a wide range of constructions, flanking conditions, and building styles. Two examples for models which fulfil these requirements are given in the next two chapters.

2.2 Statistical Energy Analysis (SEA)

Statistical energy analysis is based on a work by Lyon and Maidanik [LM62] dealing with power flow between coupled oscillators. Crocker and Price used it to calculate sound insulation of single [CP69] and double panels [CP70]. From a building acoustical point of view, it deals with power flow between reverberant sound fields be it between rooms and walls or between walls across junctions. Its basic idea is to describe sound transmission in a building (the system) in a stationary condition by the power flow between subsystems. A subsystem can be any part of the system which oscillates rather independent from other parts in the system, contains a reverberant sound field and possesses a high enough modal

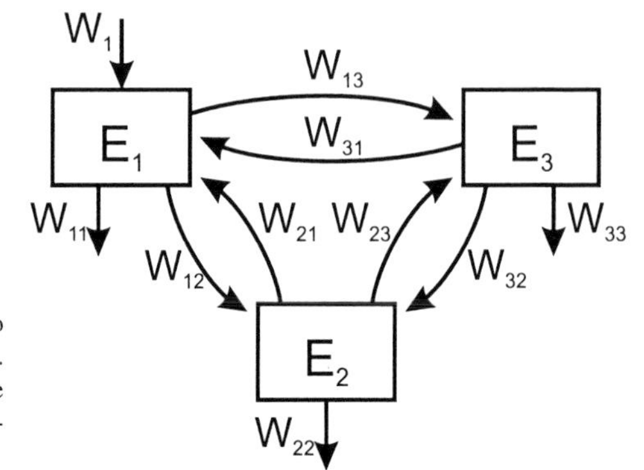

Figure 2.1: Example for SEA of two rooms with a separating wall. Subsytems 1 and 3 represent the source and receiving room, subsystem 2 the partition

density in the interesting frequency range. Different wave types are handled by different subsystems, e.g. a wall with bending, longitudinal and transverse waves would be modelled by three subsystems. When power is fed into one subsystem there will be a power loss inside this subsystem due to e.g. dissipation and an exchange of power with surrounding subsystems. These losses and exchanges are characterised by coupling loss and internal loss factors. The term 'statistical' in SEA refers to the fact that

- the energy is stored in resonant systems with high modal density. The total energy can, then, be obtained by statistical modal superposition
- the quantities are calculated in frequency bands for only the mid frequencies (average of frequency)
- one quantity is used for a large space (e.g. a room) (spatial average)
- a subsystem is characterised only by few quantities (a room e.g. by volume and absorption which means that lots of different rooms may be represented by these quantities) (ensemble average)

The mathematical model shall be explained by an example. Consider a simple room to room situation. It consists of two rooms and a separating wall. Flanking transmission and other wave types as bending waves on the separating wall shall be neglected. This situation can be rendered by three subsystems (see Figure 2.1). A source is placed in one of the rooms (subsystem 1) which injects a power W_1. Since a stationary state is considered and the energy in subsystem 1 can not grow infinitely high, there have to be power losses. The internal power loss (e.g. through dissipation) is modelled by η_1. The power flow from subsystem 1 to subsystem 2 is modelled by a coupling loss factor η_{12}. The equation for subsystem 1 yields

$$\begin{aligned} W_1 &= W_{11} + W_{12} + W_{13} - W_{21} - W_{31} \\ &= \omega\eta_1 E_1 + \omega\eta_{12} E_1 + \omega\eta_{13} E_1 - \omega\eta_{21} E_2 - \omega\eta_{31} E_3 \end{aligned} \tag{2.6}$$

where W_{21} and W_{31} denote power flows from subsystems 2 and 3 to subsystem 1. If we set the total loss factor for subsystem 1 to

$$\eta_{11} = \eta_1 + \eta_{12} + \eta_{13} \tag{2.7}$$

or in general

$$\eta_{ii} = \eta_i + \sum_{j,i\neq j} \eta_{ij} \tag{2.8}$$

the power flow in the whole system can be described by

$$\begin{pmatrix} W_1 \\ 0 \\ 0 \end{pmatrix} = \omega \begin{pmatrix} \eta_{11} & -\eta_{21} & -\eta_{31} \\ -\eta_{12} & \eta_{22} & -\eta_{32} \\ -\eta_{13} & -\eta_{23} & \eta_{33} \end{pmatrix} \begin{pmatrix} E_1 \\ E_2 \\ E_3 \end{pmatrix} \tag{2.9}$$

If subsystem 2 describes the partition between the 2 rooms, then the coupling loss factor η_{12} refers to the excitation of bending waves in the partition. It can be calculated using the consistency relationship between 2 subsystems, namely reciprocity:

$$n_i\eta_{ij} = n_j\eta_{ji} \tag{2.10}$$

where n_i and n_j describe the modal densities in the subsystems and the modal density in a room [Kut73]

$$n_{room} = \frac{4\pi f^2 V}{c^3} + \frac{\pi S f}{2c^2} + \frac{U}{2c} \tag{2.11}$$

with V: volume, S: total surface area and U: total perimeter length and c: speed of sound in air. The modal density in a wall (2D plate system) is given by

$$n_{structure} = \frac{kS}{c_g} \tag{2.12}$$

with k: wave number, S: area and c_g: bending wave group velocity.

η_{23} describes the radiation of bending waves into the receiving room and η_{13} resp. η_{31} the non-resonant transmission between the rooms due to the mass law. η_{23} can be calculated by

$$\eta_{23} = \frac{\rho c \sigma_2}{\omega \rho_{s2}} \tag{2.13}$$

where σ_2 denotes the radiation efficiency, ρ_{s2} the surface mass of the partition and ρc the characteristic impedance of air. A more detailed description of coupling loss factor expressions for different connections can be found in a paper by Craik and Smith [CS00].

To predict the insulation between source and receiving room, the coupling and total loss factors have to be known. Gibbs and Gilford [GG76] investigated the determination of these factors at different types of joints between concrete plates and the effects of simplifications as neglecting other wave types than bending waves or using the thin plate theory at low frequencies where it may not be appropriate. Craik [Cra82] made comparisons between calculations and measurements of the sound insulation of a 12 room structure which was modelled by 98 subsystems. In that early work, an accuracy of 4 dB was reached. In general, it can be said that most of the problems arise at low frequencies due to the low modal densities in the subsystems. Especially at the bending-wave coincidence frequency, higher errors arise. At high frequencies, the structural coupling loss factors are predicted somewhat too high. Craik et al. [CSE91] showed that the modes of the receiving subsystem are of greater importance for the transmission at low frequencies than the modes of the source subsystem. Special care has to be taken when double leaf lightweight constructions are to be modelled by SEA [CS00]. Different models have to be applied in different frequency ranges. The wall-cavity-wall structure can be modelled as a single wall at low frequencies but has to be modelled by 3 subsystems at high frequencies. Especially, double constructions with shallow cavities are problematic due to discrepancies between the theory of radiation into a narrow cavity. At the resonance frequency of the leafs and the cavity, the SEA predicts large dips in the transmission which do not occur in practice. Similar results are found by Elmallawany [Elm80].

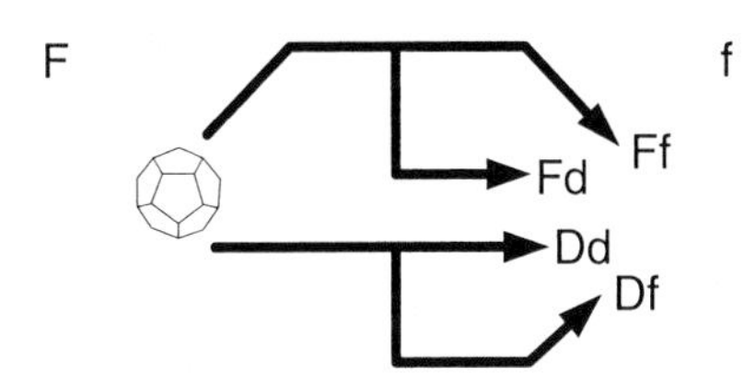

Figure 2.2: Direct and flanking transmission paths in a two-room situation Capital letters denote excitation in the source room. **Df**, e.g., means excitation of the direct partition in the source room and radiation from a flanking partition in the receiving room.

A short, not too mathematical outline of SEA and an introduction into newer methods based on SEA dealing with special problems can be found in an overview paper from Sarradj [Sar04] which some ideas for the introduction of this subsection are taken from.

2.3 EN 12354

The prediction model used in the European standard EN 12354 is based mainly on publications by Gerretsen [Ger79], [Ger86]. Contrary to the SEA method, it uses a classical approach dealing with sound pressures and velocities in diffuse fields. It can be shown, however, that it is absolutely equivalent to the calculations of a first order SEA model. It shall not be presented here in all its details since this thesis only *uses* the prediction model and does neither improve nor implement it. As much knowledge as is necessary to consider influences of simplifications, restrictions, or errors on the auralisation is given here. For more detailed information, the reader may be referred to [EN197], [Ger94] or [Met97].

The basic consideration is as follows: When the sound reduction index of a partition is measured in the laboratory, it is important to consider the loss factor since the reduction index in a real situation would be different due to vibrational energy 'escaping' through the connected structures. This could be done easily by determining the structure-borne reverberation time of the partition in the lab and in a real situation, $T_{s,lab}$ and $T_{s,situ}$. Furthermore, the vibrational energy transmitted through the flanking paths will itself contribute a part to the sound power in the receiving room and has to be considered. Therefore the transmission via connected structures has to be quantified by material and construction parameters. In general, it can be said that the sound power in the receiving room can be calculated by adding the power transmitted via different independent paths that are the direct path via the separating wall and flanking paths (see Figure 2.2).

Starting from the radiated power in the receiving room

$$W_r = \rho c v^2 S \sigma \tag{2.14}$$

with v denoting the particle velocity of the wall surface, S and σ the area and density of the partition and considering τ deduced from the sound reduction index

$$\tau = \frac{W_r}{W_s} = \frac{p_r^2}{p_s^2}\frac{S}{A} \tag{2.15}$$

with p_s and p_r denoting the sound pressures in source and receiving room and A denoting the equivalent absorption area, resulting in

$$W_r = \frac{p_s^2 A}{4\rho c}\tau \tag{2.16}$$

one obtains

$$v^2 = \frac{p_s^2}{4\rho^2 c^2}\frac{\tau}{\sigma} \tag{2.17}$$

Using a definition for the vibration reduction factor

$$d_{ij} = \frac{v_j^2}{v_i^2} \tag{2.18}$$

the sound pressure in the receiving room due to transmission via a path from element i in the source room to element j in the receiving room (see Figure 2.2) becomes

$$p_{r,ij}^2 = p_s^2 \tau_i d_{ij} \frac{\sigma_j}{\sigma_i}\frac{S_j}{A} \tag{2.19}$$

and, thus, an expression for the flanking transmission coefficient can be found as

$$\tau_{ij} = \frac{p_{r,ij}^2}{p_s^2}\frac{A}{S_0} = \tau_i d_{ij}\frac{\sigma_j}{\sigma_i}\frac{S_j}{S_0} \tag{2.20}$$

with the corresponding sound power reduction index for path ij

$$R_{ij} = R_i + D_{v,ij} + 10\log\frac{S_j}{S_0} + 10\log\frac{\sigma_j}{\sigma_i} \tag{2.21}$$

Using the reciprocity of the flanking transmission coefficient $\tau_{ij} = \tau_{ji}$, the radiation factors σ_i and σ_j which mostly are not known and hard to determine correctly can be removed. This yields

$$R_{ij} = \frac{R_i + R_j}{2} + \overline{D_{v,ij}} + 10\log\frac{S_0}{\sqrt{S_i S_j}} \tag{2.22}$$

with

$$\overline{D_{v,ij}} = \frac{D_{v,ij} + D_{v,ji}}{2} \tag{2.23}$$

Equation (2.22) contains either known or easy to determine quantities except for the vibration level difference $\overline{D_{v,ij}}$. This quantity can be determined by measurements but since it is not invariant related to different field situations, i.e. it will give different results in different situations, it is desired to use an invariant quantity to consider the vibration transmission across junctions. This can be reached by referencing the differences between situations to a fixed quantity to make it a parameter describing the 'material' itself. Consider the sound reduction index R: If the room volume or the size of the partition is changed, R stays constant due to the relation to the absorption area and the partition area A/S. The same principle is used when the vibration reduction index

$$K_{ij} = \frac{D_{v,ij} + D_{v,ji}}{2} + 10\log\frac{l_{ij}}{\sqrt{a_i a_j}} \tag{2.24}$$

is introduced since it is the vibration level difference across a junction referred to the length l_{ij} of the junction between elements i and j and the equivalent absorption lengths a_i and a_j of the elements. Last terms can be expressed by

$$a_i = \frac{2.2\pi^2 S_i}{cT_{S,i}}\sqrt{\frac{f_{ref}}{f}} \tag{2.25}$$

and denote the length of a junction in meters with a transmission coefficient for bending waves of 1 which cause the same losses as the existing construction. f_{ref} is set to 1000 Hz.

Finally, for the airborne sound insulation of a room situation, the apparent sound reduction index

$$R' = -10\log\left(10^{-R_{Dd}/10} + \sum 10^{-R_{ij}/10}\right) \tag{2.26}$$

is obtained.

Impact sound insulation For calculation of impact sound insulation, the impact sound levels of the products are used:

$$L'_n = 10 \log \left(10^{L_{n,d}/10} + \sum 10^{L_{n,ij}/10} \right) \tag{2.27}$$

Similar to the deduction of the equations for airborne sound insulation,

$$L_{n,ij} = L_{n,i} + \frac{R_i - R_j}{2} - K_{ij} + 10 \log \frac{l_{ij}}{l_0} - 10 \log \frac{S_i}{S_0} \tag{2.28}$$

can be deduced with $l_0 = 1$ m and $S_0 = 1$ m^2.

Additionally to the detailed model as described above, the EN12354 also specifies a simplified model. For this, only single numbers according to ISO717 without corrections regarding to the structural reverberation time T_s are used. Since for auralisation the frequency-dependent quantities have to be known, the simplified model is not of great interest in this thesis.

Restrictions / Assumptions For the detailed and the simplified model, the following restrictions resp. assumptions have to be mentioned:

- The calculation models may only be applied to adjoining rooms.
- Only transmission along one junction is considered. The rear wall in a horizontal room situation, e.g., is not considered.
- For room situations with big-sized stiff partitions and flanking partitions in the receiving room with a low stiffness, the flanking transmission is calculated too high.
- The radiation on both sides of a partition is considered as equal.
- Only combinations of products are suitable which the vibration reduction factor is known for.
- The simplified model is only valid for room situations with dimensions similar to those in testing facilities.

- The impact sound reduction improvement ΔL measured on a heavy-weight construction according to DIN EN ISO 140-8 can not be applied to light-weight constructions.
- The simplified model for impact sound insulation can only be applied to single layered massive floors with floating floors or coverings with dimensions as in common buildings.

Furthermore, problems arise with the calculation of

- flanking partitions with orifices
- staggered joints
- different materials on both sides of the joints

which, however, can be coped with by correction terms and simple rules of thumb [BFSS02].

Verification During the development of the calculation model, lots of field measurements have been used to find relations and correction terms. Thus, it can be expected, that the accuracy of the model is quite good. In the past, several comparisons between measurements in buildings and calculations according to DIN EN 12354 have been carried out by Fischer and colleagues [SBF01], [BSFS01], [BFSS02], [SKF04].

Späh et al. [SBF01] considered the influence of input data like loss factors, vibration reduction factors for different junctions etc. which are given for different materials in supplements to EN12354 and can be used if no measured data is available. It was found that the accuracy of some of the input data could be improved. Subsequent to this, Blessing et al. [BSFS01] investigated the accuracy of the calculation model by comparison with measurements. Different calculation software was used (Bastian, Acoubat, self-made programs) and it was found that

- the insulation is predicted as too low if only input data from the supplements of EN 12354 are used. The mean deviation is 2.2 dB.

- Acoubat predicts values which are 0.6 dB below the measured values.
- calculations with Bastian predicts sound insulation 1.6 dB too high.
- for the improved input data from [SBF01] a mean deviation of 0.3 dB could be reached.
- the standard deviation for all versions of the calculation is about 2.5 dB.

Measurements on massive constructions have shown that the vibration reduction index is relatively constant over frequency except for frequencies below 200 Hz and above 1250 Hz where K_{ij} is rising. This is due to low modal densities in the lower frequencies and a decay over propagation in the wall at high frequencies.

In a comparison of measurements and calculations on 31 building situations with mainly heavy and separating and flanking elements, Metzen [Met99] found a deviation of (2.0 ± 1.8)dB for the simplified model, (2.1 ± 1.8)dB for the detailed model and (5.0 ± 1.8)dB for the detailed model with correction of the structural reverberation time (the values denote the mean deviation $\pm$ the standard deviation).

After all, it must be stated that the calculation model yields satisfactory accuracy and is surely a signiificant extension of the existing standard rules.

Chapter 3

Principles of Auralisation

When the quantities for sound insulation have been calculated as described in the previous chapter, auralisation comes into play. Its aim is to calculate the sound pressure in the receiving room at the ears of the listener and replay the signals by an appropriate equipment. In this chapter, the tools and methods as well as the hardware needed will be described. In section 3.1 the basic signal processing algorithms are presented. Section 3.2 describes the principles for doing binaural auralisations such as HRTFs, dummy head recordings, etc. and at last in 3.3, the basics of sound reproduction, level calibration, equalisation and dynamics will be expounded.

3.1 Signal Processing

Any analog signal, be it sound pressure, velocity, or force has to be transformed into a series of numbers to be processed by a computer. The process of sampling measures the quantities - normally voltages from a transducer - at discrete times and transforms them into numbers of a certain range which is determined by the numbers of bits used to represent the numbers. The representation of such a digitised signal, normally, is given as $s[k]$ in the literature where k denotes the number of the sample. Together with the sampling rate, the time value can be calculated. $s(k)$ may be the recorded signal of a piano, a machine, or a passing train but may also be the force signal of a tapping machine or a walking person.

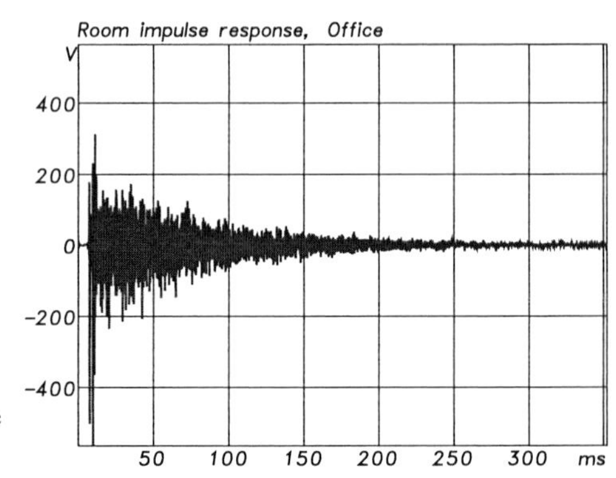

Figure 3.1: Room impulse response of an office room

Impulse Responses The propagation of a signal through a room, across a structure, etc. can be expressed by an impulse response of the system. Consider a source and a receiver inside a room. The sound at the receiver is modified by the characteristics of the room like reflections and damping. This can be described by an impulse response $h(k)$ of the room which is the signal at the receiver if the source emits a single pulse. In the simple case where the room only consists of one wall (the other ones are totally absorbent), the impulse response consists of two pulses - the direct sound and a reflection of the wall which travels a longer distance and arrives later at the receiver and has a smaller amplitude due to energy loss from the reflection. In Figure 3.1 the impulse response of an office room can be seen.

Convolution From an input time signal $s(k)$ and an impulse response of any linear, time invariant (LTI) system $h(k)$, the time signal at the output of the system $g(k)$ can be calculated. Thus, e.g., the time signal at a receiver in a room can be calculated from the source signal and the impulse response of the room. The operation can mathematically be described by

$$
\begin{aligned}
g(k) &= \sum_{\nu=-\infty}^{\infty} s(k)h(k-\nu)d\nu \\
&= s(k) * h(k) \\
&\quad ⫯ \\
G(f) &= S(f) \cdot H(f)
\end{aligned}
\tag{3.1}
$$

in the time and frequency domain. Convolution is the basis for auralisation and can be carried out in the time or frequency domain. For longer impulse responses, the convolution is carried out in the frequency domain for sake of processing speed.

Interpolation The input quantity for auralisation, normally, is given in one-third octaves or even in octaves which range from 50 Hz or 100 Hz up to 3150 Hz or 5000 Hz. For signal processing, these quantities have to be turned into frequency spectra with an expediant number of frequency lines. In the following, an input signal with 21 values in one-third octave bands ranging from 50 Hz to 5000 Hz is assumed. To obtain a frequency spectrum with 4097 lines, these values have to be interpolated. This can be done in different ways. The easiest way is to fill all frequency lines in a band with the value at the mid-frequency. More sophisticated algorithms use linear interpolation, continuous averaging, or cubic spline interpolation. Examples of these methods can be seen in Figure 3.2.

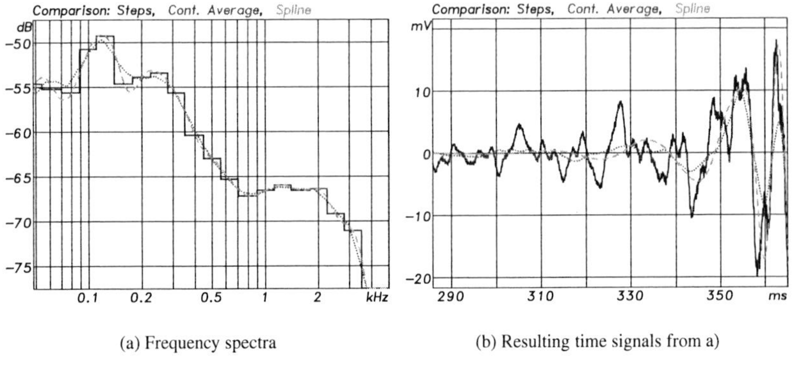

(a) Frequency spectra

(b) Resulting time signals from a)

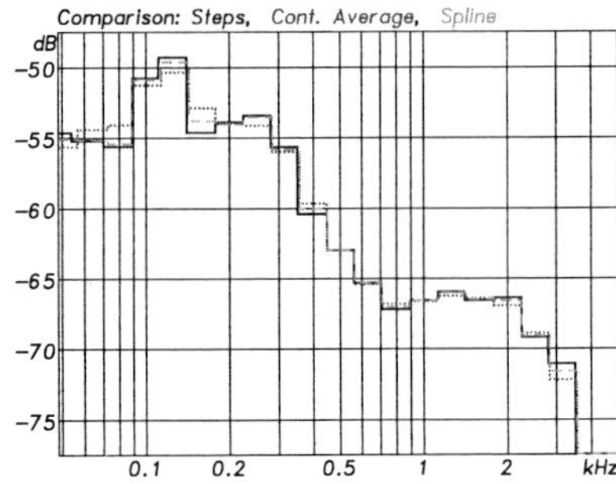

(c) Original step function and averaged interpolated functions

Figure 3.2: Different methods for interpolation in the frequency domain (a), influences on the time signal (b), and re-averaging (c). Figure a shows the frequency spectra for a step function (blue, solid), continuous averaging of the step function (red, dotted), and an interpolation with cubic splines (green, dashed). In Figure b the corresponding time signals can be seen. It is obvious that the step function has the worst transient behaviour. Figure c shows the results of averaging the functions in one-third octave bands. The result should be identical to the original one-third octave band data (blue, solid). The spline interpolation shows the best behaviour and is, therefore, used in this thesis.

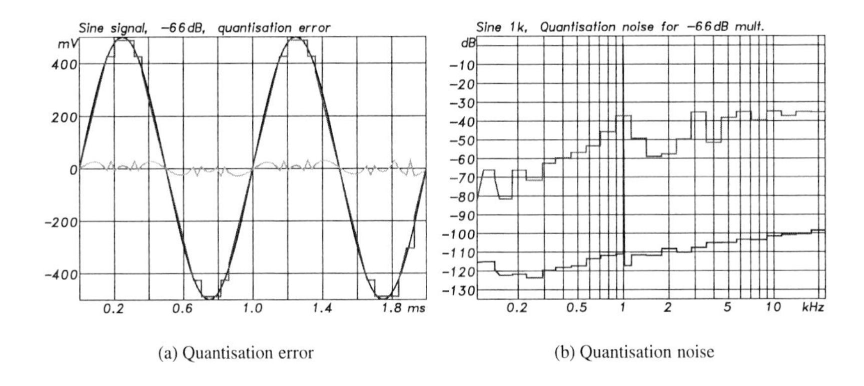

(a) Quantisation error

(b) Quantisation noise

Figure 3.3: Quantisation noise and reduced dynamics. Figure a shows a 1 kHz sine signal quantised with 16 bits which is scaled down by -66 dB and up again by 66 dB. The deviation from the original signal due to quantisation error can be seen by the difference signal. Figure b shows the corresponding frequency spectra (added averages). The upper curve represents the quantisation noise introduced by the multiplications. The dynamics are reduced by app. 60 dB.

Quantisation and Dynamics In a computer, signals are represented as numbers. After sampling, the samples are represented by fixed point numbers with typically 16 bit resolution. To avoid quantisation errors (see Figure 3.3), floating point arithmetics are used where the numbers are divided into an exponent and a mantissa. The mantissa is always shifted in a way that all possible bits are used which reduces the quantisation error to a minimum. All signal processing operations in this work are using floating point numbers. Disadvantages of this are the increased memory usage which is usually 4 times higher and the loss of processing time by conversions between fixed and floating point representation. The advantage is the higher dynamics, that is the level difference between the signal peak and noise is increased. A simple rule of thumb is that with every bit used for fixed point numbers, the dynamic range is increased by 6 dB. Thus, a CD with its 16 bit resolution reaches a dynamic range of app. 96 dB. This means that if the volume of a PA-system is set to a peak level of 130 dB, the quantisation noise would be audible at a level of 34 dB.

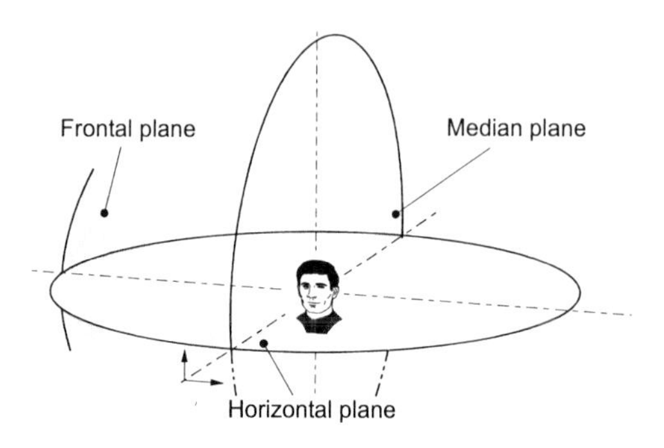

Figure 3.4: Planes in binaural hearing

Another point of concern is the dynamic range of the playback system. This is covered in section 3.3.

3.2 Binaural Techniques

Humans, normally, use two ears for listening. Due to the fact that we intend to create a spacial impression of listening in a room, it is, therefore, necessary to consider an auralisation with binaural signals. This means that the influence of the shape of head and torso on hearing must be considered. This section discusses the implications of creating binaural signals.

Due to the shape of head and torso, information is added to incoming sound signals which allow the user, e.g., to locate sound sources. If a sound originates from the left of a person, it arrives at the left ear first. Also, there is a difference in level between the ears due to the shadowing effect of the head. For sources in the median plane, none of these effects occurs. Here, it is possible to locate the sound due to the non-symmetrical shape of the head. Sounds from the front and the back, e.g., can be distinguished by their different frequency spectra.

These time and frequency characteristics have to be put into the signals and impulse responses during simulation.

The influences of head and torso can be described as an LTI system and, thus, by head related impulse responses (HRIR) or the corresponding head related transfer functions (HRTF). These can be measured for dummy heads (see Figure 3.5 a) and individuals. Ideally, for each listener, his own HRTFs are used since there are differences between the individual heads. This, however, is nearly impossible in practice. Therefore, either lots of measurements on individuals have to be carried out and an average has to be taken or a dummy head with an average shape is used (see [Gen84]). In the following, the HRTFs of the ITA dummy head are used [Sch95]. For the measurement, a loudspeaker is moved around the dummy head and the transfer functions from the speaker to the microphones in the ear canal of the head are measured for the desired positions or angles. Figure 3.6 shows results of these measurements for a source in 50° to the front left of the head. If a monophonic signal of a source is convolved with the HRIR, the source is virtually shifted to the direction from which the HRIR was measured. Thus, if the direction which the sound comes from is known, the binaural cues can be easily included by applying the appropriate HRIR. To describe the influence of the head in free and diffuse field conditions, the free field and diffuse field transfer functions are used. The first one is just the HRTF from the front, the latter one is the energetic sum of all HRTFS. These are shown in Figure 3.5 b and c.

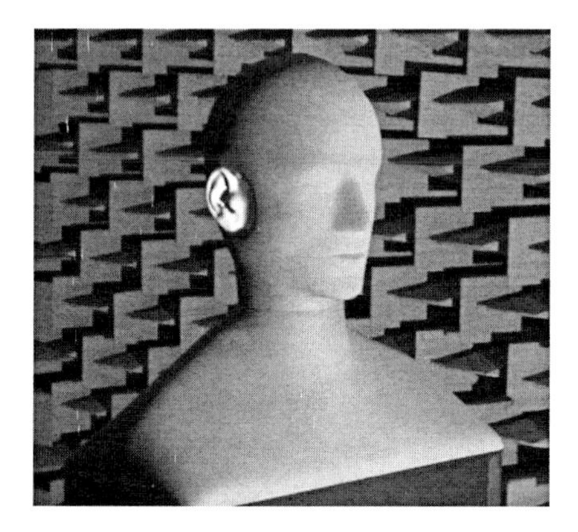

(a) Dummy head

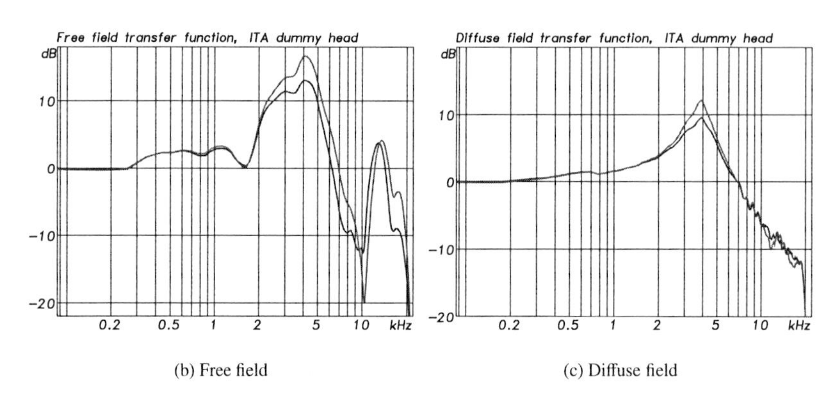

(b) Free field

(c) Diffuse field

Figure 3.5: Free field and diffuse field transfer function of the ITA dummy head.

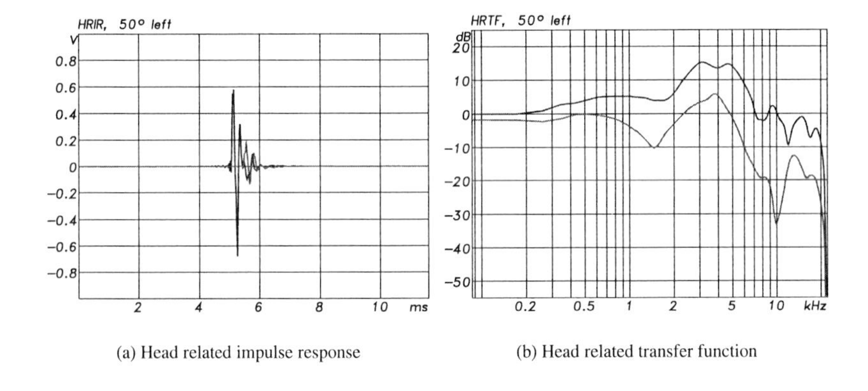

(a) Head related impulse response

(b) Head related transfer function

Figure 3.6: Head related impulse response (a) and the corresponding transfer function (b) for an azimuth of 50° and an elevation of 0° (front left). The left channel of the time signal starts earlier and has a higher amplitude. Also the left frequency spectrum has a higher amplitude and a different shape. The increased level between 2 kHz and 6 kHz is typical for HRTFs and can be related to the ear canal resonance.

Binaural room impulse responses consist of an impulse response for each ear and can be measured with dummy heads. Figure 3.7 shows the structure of an impulse response and the geometric relations inside a room.

If a dry source signal is convolved with, e.g., a binaural room impulse response measured in the dome of cologne and replayed, the listener hears the same sound he would hear if he stood in the place where the impulse response was measured. In general, when listening to binaural signals, no crosstalk between the signals for the left and right ear must occur. Otherwise the reproduction is not true and the '3-D effect' is lost. Replaying the signals by headphones is, normally, no problem but when loudspeakers are used, the left ear signal also gets to the right ear of the listener. In some cases, it is desirable not to use headphones because of the unnatural feeling, in-head localisation of sounds originating from the median plane, etc. For reproduction by loudspeakers, it is possible to cancel the crosstalk between the channels. This means, the listener is positioned in front of two

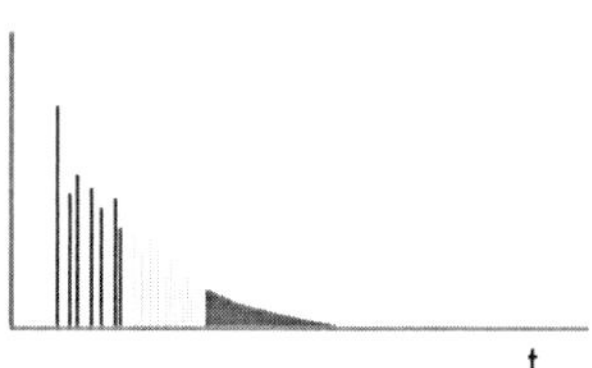

Figure 3.7: Room with a source and a receiver and the composition of the energy impulse response (for one ear). Blue: direct sound, red: first order reflections, yellow: second order reflections, pink: diffuse reverberation. From [AE03]

loudspeakers and due to appropriate signal processing, the signal from the left speaker to the right ear and vice versa is cancelled. With adaptive systems, this does not only work for one spot but also for moving listeners.

In the algorithms for auralisation of airborne and impact sound insulation which are developed in this thesis, measured binaural room impulse responses are used for modelling reverberation. This is explained more detailed in Chapters 4 and 5.

3.3 Presentation Techniques

The presentation of signals resulting from building acoustical auralisation differs in an important point from other techniques: The relevance of loudness. When listening to room acoustical auralisations, the colouration, the spaciousness or lateral fraction are important which do not change much with level. In building acoustical auralisation, the level is the most important thing together with colouration. Therefore, care has to be taken for the

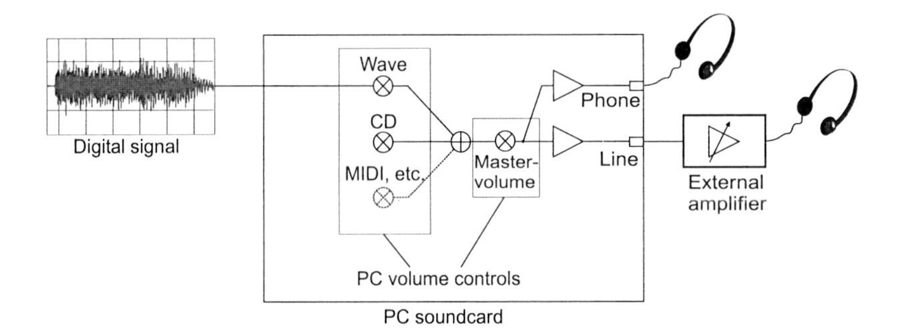

Figure 3.8: Signal flow and controls for audio hardware for auralisation

reproduction of both the correct absolute level and the relative level between source and receiving room. Since some room situations have level differences of 50 dB and more, care has to be taken for not wasting valuable signal to noise ratio in the signal chain. This section discusses the points of concern when auralised signals have to be replayed like dynamic ranges of soundcards and amplifiers, different types of headphones, etc.

The typical scheme for the sound reproduction hardware is shown in Figure 3.8. The digital signal represented by numbers in memory is replayed via a soundcard with multiple software volume controls. A headphone can be directly connected to the phone output of the card or via an external amplifier, e.g. a hifi amplifier, to the line output of the soundcard which adds one more volume control to the chain. In usual soundcards, the different virtual volume controls, normally, correspond to only one hardware or software control for sake of lower costs. Thus, it does not matter which volume control is used.

The situation shall be explained by an example: Consider two sound signals: A circular saw with a level of $L_{A,F,max}$ = 105 dB(A) and a vacuum cleaner with $L_{A,F,max}$ = 75 dB(A). The room situation to be auralised has a sound reduction index of R'_w = 53 dB. Figure 3.9 shows the time averaged sound pressure levels in one third octave bands. The upper curves represent the level of the source room signals; the lower curves represent the receiving room signals which are clearly lower in level due to the sound insulation. It can be seen

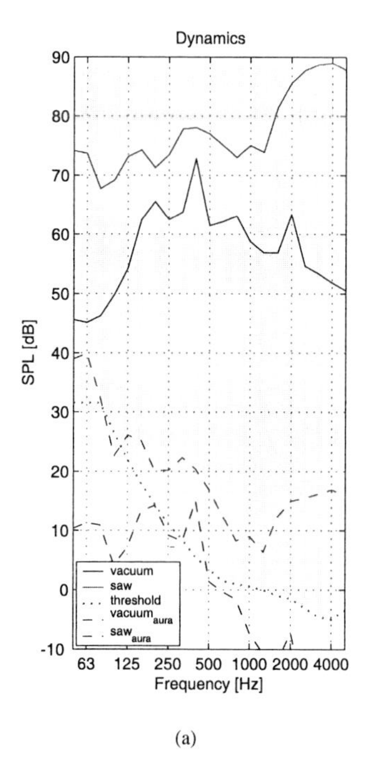

(a)

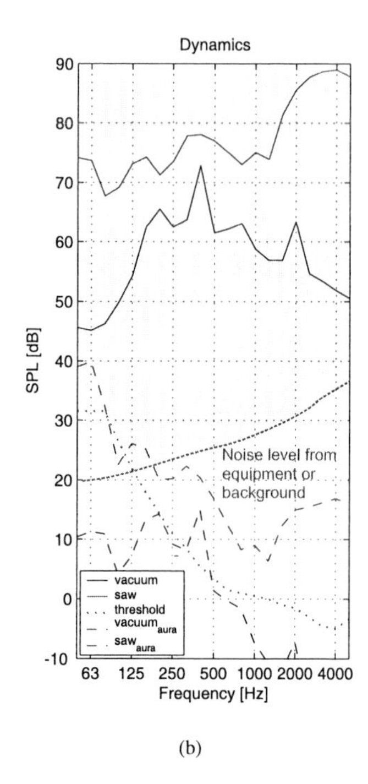

(b)

Figure 3.9: Considering the dynamical range: Levels in the source room (solid curves in blue and black), in the receiving room (dashed curves in blue and black) and the hearing threshold (red, dotted). The grey shaded area is the dynamical range the equipment must provide. Figure a shows the ideal situation with no background or equipment noise. Figure b represents a real situation with noise either from the equipment (soundcard, external amplifier) or from the background (ventilation systems, traffic noise from outdoors, etc.)

that most of the lower signal (vacuum cleaner) falls below the threshold of hearing which is the lower border for the necessary dynamics. The circular saw is higher in level and marks the upper border of the dynamical range. The range to be covered by the equipment is shown by the grey shaded area. It has to be considered that the levels shown are mean values calculated from the entire signal. The upper level, however, is determined by the maximum peak of the signal. As the $L_{A,F,max}$ of the circular saw is at 105 dB(A), the range must either be extended to this value or the signal has to be processed in a way that the peak values are smoothened out. It can, thus, be seen that the dynamical range of the equipment must be about 105 dB for this example if parts of the signal which are just below the hearing threshold should be hearable. In normal listening situations, however, the background noise is about 30-35 dB and in ambitioned listening tests, maybe, a background level of 15-20 dB can be reached. The dynamical range is, then, reduced to about 90 dB which is still a value being found only on higher quality soundcards. The maximum range achievable with a 16 bit A/D converter is 96 dB. Usual soundcards, often built into the mainboard of a PC or a laptop deliver a dynamical range of 60-80 dB. If the absolute level of the sound signals are to be reproduced, a calibration of the replay chain has to be done. This can be achieved by replaying a sound signal which marks a point in the hearing threshold, for example, a noise signal with a bandwidth of one-third octave at 80 Hz. The hearing threshold for 80 Hz is at 30 dB. If the user adjusts the volume control in a way that the sound is just audible, the system is calibrated since all replayed sounds can be referenced to the calibration signal.

Headphone Equalisation Normally, for presentation of binaural signals, headphones are used since they are easy to handle and provide the necessary separation of the left and right ear signal. It is, however, problematic to use different types of headphones since they have different transfer functions. Figure3.10 shows transfer functions of different headphones. It can be seen that they differ in level and frequency shape. Ideally, the transfer functions have to be equalised for reproduction which is, however, not possible if auralisation is used in the field and not in research. For the comparison of different building situations, however, this fact is not that problematic since all signals are listened to by the same headphone and the relative levels are correct.

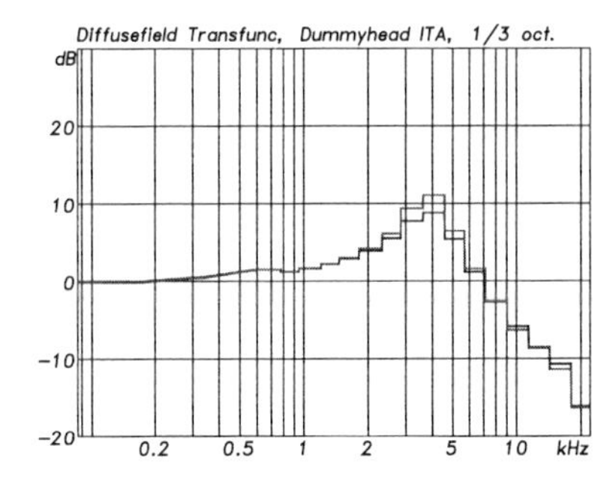

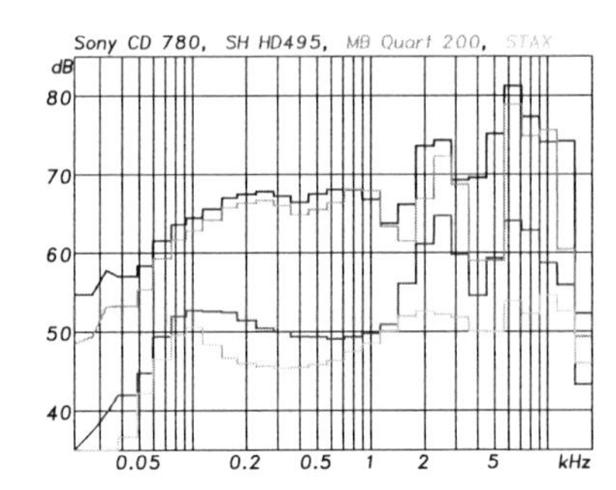

Figure 3.10: Diffuse field transfer function of the ITA dummy head and the diffuse field transfer function of 4 headphones referenced to the same input voltage. The STAX headphone has an amplifier in the measurement chain, so it is not comparable in level

Cross-talk Cancellation Presentation of binaural signals by headphones has the drawbacks that people feel uncomfortable and unnatural wearing a headphone and in-head localisation occurs which means that the sound source seems to come from inside the head. If loudspeakers are used for presentation of binaural signals, the problem arises that the channels for the left and right ear are not separated anymore. Consider a pair of loudspeakers 45° in front of a listener. The signal from the left channel which is determined only for the left ear also reaches the right ear. It is, however, a little lower in level due to the shadowing effect of the head. Due to this cross-talk, the effect of the binaural signal to be 3-dimensional is destroyed. Sound sources, then, can only be localised between the angle spanned by the speaker system.

To cope with this, it is necessary to cancel the cross-talk. This can be reached by replaying cancellation signals which interfere with the undesired signals at the ears in a way that they are not audible. This effect can be reached by applying FIR filters which reduce the cross-talk between the ear signals to a reasonable amount. A detailed description of this technique can be found in [Sch93] and [LS02].

The resulting system can be considered as an open headphone though the drawbacks of headphones are compensated.

Chapter 4

Filters for Airborne Sound Insulation

This chapter describes the algorithm for auralisation of airborne sound insulation. It was first published in 2000 [VT00a], [VT00b]. In summary, from the standardised level difference D_{nT} of the direct and flanking transmission paths and the characteristics of the receiving room, the sound pressure signal at the listener's ears in the receiving room is calculated and can, then, be replayed by a standard sound card.

To get an overview of the physical mechanisms and the important parameters to be modelled, the process of sound transmission between two rooms shall be described. In the source room, a sound field with direct sound and reverberation is produced by a source. Airborne sound is transformed into structural waves on the walls (particularly bending waves) which are transmitted through the building structure to the receiving room. Airborne sound is radiated from the walls and propagates to the receiver. An exact model of this process would include a room acoustical simulation of the sound field in the source room, the modelling of the excitation and radiation of bending waves in the walls, and a room acoustical simulation of the sound field in the receiving room. On the other hand, the important parameters for the perception of sound in the receiving room must be identified. Contrary to room acoustical auralisations, where parameters like spaciousness, interaural cross correlation (IACC), lateral fraction etc. are important to judge the acoustical quality of a room, in building acoustical auralisation, the level, the colouration and a realistic binaural impression are the most important parameters. It is questionable if an exact modelling is necessary to reproduce these parameters. Thus, simplifications can be introduced which significantly reduce the complexity and required calculation time of the simulation. To reproduce level and colouration, it would be sufficient to model the sound transmission by the level difference between source and receiving room. This could be done by a simple filter function (e.g. a graphic equalizer in which the level differences are set to the appropriate frequency bands). This approach, however, neglects the influence of the reverberation in the receiving room and also no directional information is included which is important for a natural reproduction of the sound field. In this work, reverberation and binaural cues are considered by using binaural room impulse responses and binaural impulse responses and the corresponding HRTFs of the ITA dummy head [Sch95]. A detailed description of the model and possible errors due to simplifications follows below.

The Model For the auralisation model, input data are required. It is important to keep the amount of input data as low as possible and close to the customary parameters used to describe sound insulation to cover a wide range of applications. The input data are, normally, delivered by prediction models as described in Chapter 2. Thus, the approach is derived from the situation that the input data are the standardised level difference D_{nT} and geometrical information of the room situation.

A building acoustical measurement of the standardised sound level difference is described by

$$D_{nT} = L_S - L_R + 10\log\frac{T}{0.5\text{ s}} = D + 10\log\frac{T}{0.5\text{ s}} \tag{4.1}$$

with L_S denoting the average level in the source room, L_R the average level in the receiving room, and T the reverberation time in the receiving room. The level difference D is 'responsible' for level and colouration. The reverberation is described by T. Equation (4.1) can also be written in scale of squared sound pressures as

$$\frac{1}{\tau_{nT}} = 10^{\frac{D_{nT}}{10}} = \frac{p_S^2}{p_R^2}\frac{T}{0.5\text{ s}} \tag{4.2}$$

with p_S and p_R denoting the sound pressures in the source and the receiving room and τ_{nT} denoting the standardised transmission coefficient. For the purpose of auralisation, equation (4.2) is rearranged. The receiving room sound pressure in one-third octave bands, thus, can be derived from

$$p_R^2 = p_S^2 \tau_{nT} \frac{T}{0.5\ \mathrm{s}} \tag{4.3}$$

It should be noted that τ_{nT} is composed of the sum of all transmission paths. In terms of sound pressure signals flowing through an acoustic linear and time invariant system in the entire frequency range, the equation reads

$$p_R(\omega) = p_S(\omega) \sum_{j=1}^{5} f_{\tau,i}(\omega) f_{rev,j}(\omega) \tag{4.4}$$

with $f_{\tau,j}$ denoting filter functions related to the transfer functions between the source room and the radiating walls in the receiving room. $f_{rev,j}$ is the transfer function between the radiating wall and the receiver. This equation is at least correct for certain simplifications (see below). Except for the phases, the filters can be identified unambiguously. $f_{\tau,j}$ must have exactly the same one-third octave band spectrum as the corresponding path transmission coefficient, and $f_{rev,j}$ is a classical room transfer function derived from the impulse response between the wall and the receiver.

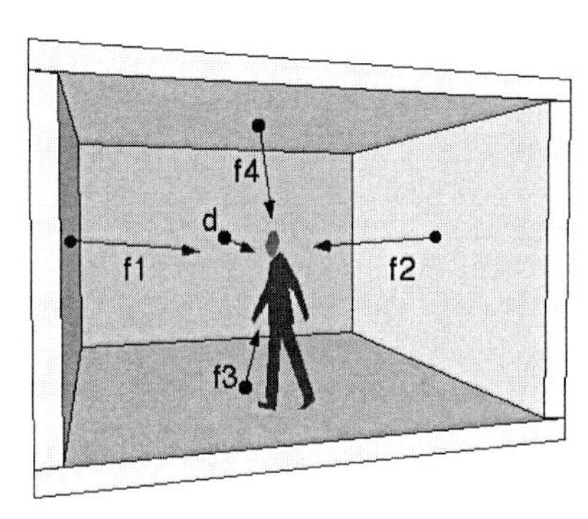

Figure 4.1: Situation in the receiving room with listener and 5 point sources in the centre of each wall

To make this approach vivid, the situation in the receiving room as in Figure 4.1 is considered: A listener in the receiving room is standing at the central position of the room, 10% moved in the direction of and facing the separating wall. Sound is propagating from the source room to the receiving room and radiated from the direct and flanking partitions. Since the input data for the auralisation comes mostly from the EN12354 calculation algorithm, only first order junctions are considered. Thus, the back wall is not radiating any sound in this example. The radiating walls are modelled as point sources located in the middle of each wall[1]. Since the geometric data of the room and the position of the listener is known, the distances and directions of the point sources relative to the listener can be calculated. Inside the receiving room, the direct sound radiated from the walls can be considered as 5 independent point sources. The original source is located in the source room, thus, the sound spectra of the 5 sources in the receiving room are frequency shaped by the level difference between source and receiving room.

[1] It is shown in listening tests that no improvement can be found using surface sources [Hol02]

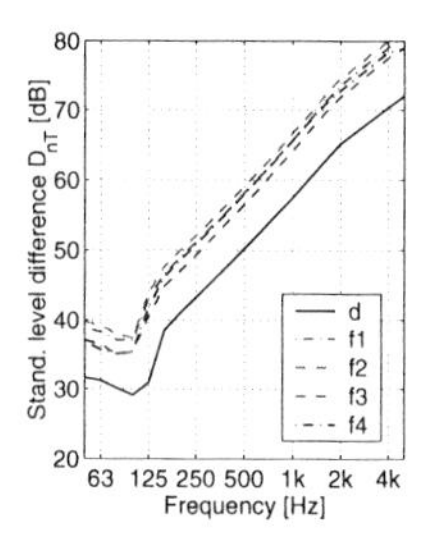

(a) Input data D_{nT}

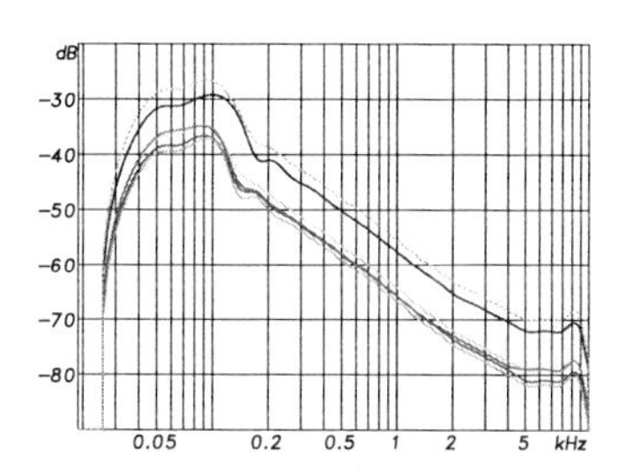

(b) Interpolated spectra, dashed line represents the energetic sum of the transmission index $\sum \tau_i$

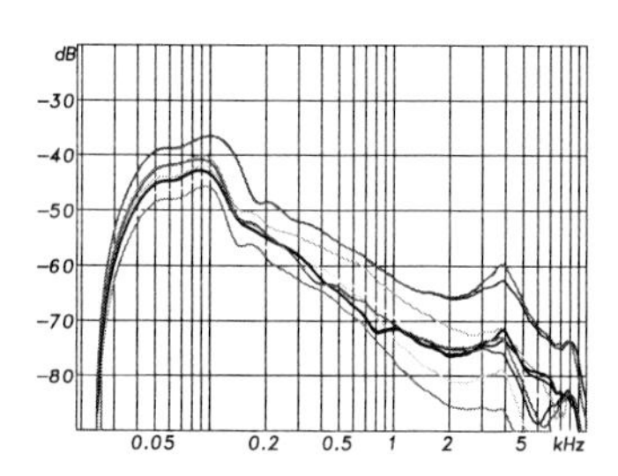

(c) Amplitude weighting and inclusion of HRTFs. Binaural signals.

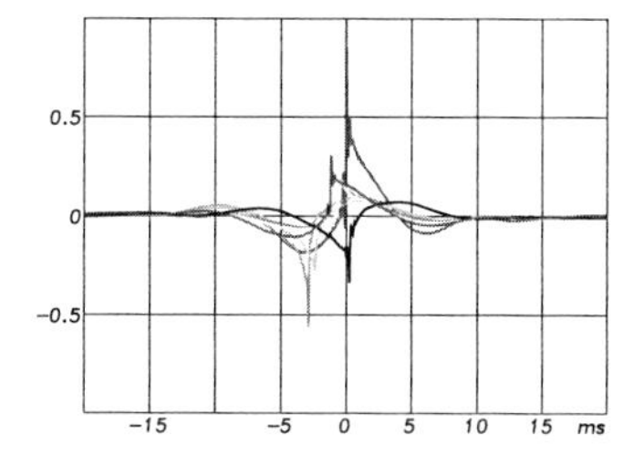

(d) 5 binaural direct sounds

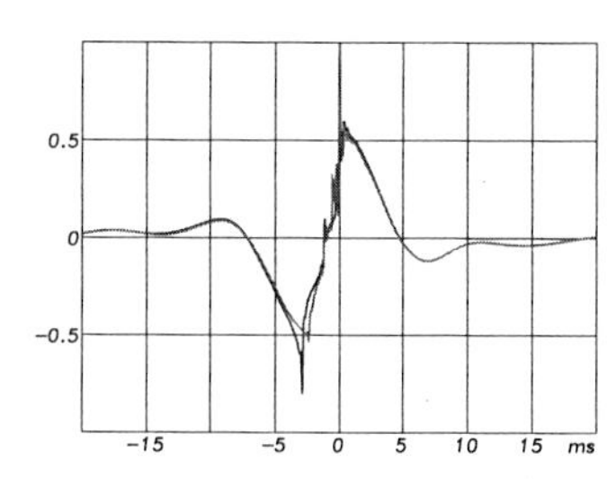

(e) Sum: Direct sound cluster

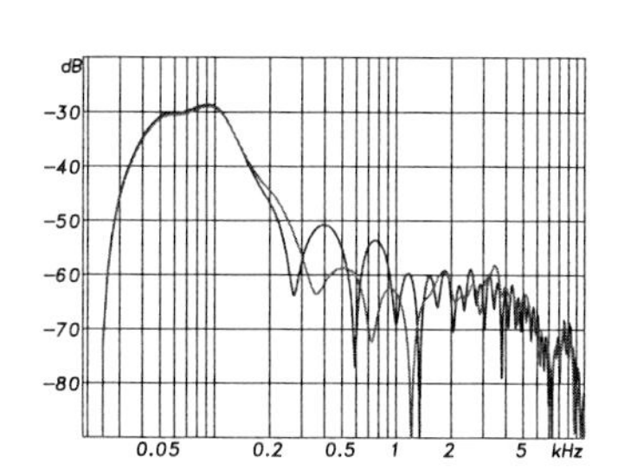

(f) Spectrum of the direct sound cluster

Figure 4.2: Steps for generating the direct sound cluster

Figures 4.2 a - f show the steps necessary for generating the direct sound part of the impulse response of the transmission from the source to the receiving room. As input, the values of the standardised level difference D_{nT} (Figure 4.2 a) are used. In a first step, the one-third octave band data are processed to obtain a spectrum with 4097 frequency lines (see section 3.1, page 27). Figure 4.2 b shows the result of this operation. These spectra represent acoustic filters for the transmission of sound from the source room to the receiving room. The values represent the transmission coefficients τ_i in a logarithmic scale referenced to 1 and are negative, since the sound is attenuated when transmitted through the partitions. The input data, normally, are present only for frequencies between 100 Hz and 3150 Hz so an appropriate handling for the frequencies below and above these values must be applied. It was chosen to apply slopes with -36 dB per octave. The range of the original input data is marked by the vertical lines in the figure. Since no information about the filter phases is present, a linear phase is applied. Then, the distance between the radiating walls and the listener is considered by amplitude weighting and time delay. The direction of the sources relative to the listener are considered by applying the head related transfer functions between the points of radiation on the walls and the listener. The result can be seen in Figures 4.2 c (spectra) and d (corresponding time signals). Afterwards the direct sounds are added to form the direct sound cluster of the impulse response p_{direct} (Figure 4.2 e). Figure 4.2 f shows the corresponding spectrum.

In the receiving room, a reverberant sound field is excited by the transmitted direct sounds. To model this sound field, a measured binaural room impulse response is used. As a simplification, only 1 reverberation process for all 5 point sources is considered. It is assumed that the reverberant part of the binaural room impulse response does not change significantly for different source receiver positions. Thus, using only one reverberation process instead of 5 does not introduce a significant effect. Consider the situation in the receiving room: The listener is positioned near the center of the room. In a usual reverberant room (0.4 s $\leq T \leq$ 1.5 s) the diffuse field distance where the energy of the direct field and the diffuse field are equal is always smaller than the distance between the source and the listener. This means that the energy of the diffuse field is significantly higher than the direct

Room	V [m^3]	T(500 Hz) [s]	A [m^2]	r_H [m]	$d_{S,R}$ [m]	$\frac{E_{Diffuse}}{E_{Direct}}$	$\frac{E_{Direct}}{E_{Total}}$
Bathroom	19	0.6	5.6	0.33	1	8.9	0.10
Bathroom, unfurnished	26	1.0	4.2	0.29	1	11.9	0.08
Office room	57	0.5	18.6	0.61	1.5	6.1	0.14
Kitchen, unfinished	26	1.0	4.2	0.29	1	11.9	0.08
Sleeping room	38	0.4	14.8	0.54	1	3.4	0.23
Seminar room	95	0.5	29.2	0.76	2	6.9	0.13
Lecture room	370	1.2	50.3	1.00	3	9.0	0.10
Living room	78	0.6	21.9	0.66	2	9.2	0.10
Living room, unfurnished	141	3.1	7.4	0.38	3	61.0	0.02

Table 4.1: Room acoustic quantities for 9 example rooms. The quantities denote the Volume V, the reverberation time for 500 Hz T, the equivalent absorption area A, the diffuse field distance r_H, the distance between source and receiver $d_{S,R}$ if the listener is positioned near the center of the room, the relation between diffuse field energy and direct sound energy $\frac{E_{Diffuse}}{E_{Direct}}$, and the relation between direct sound energy and the total energy of the impulse response $\frac{E_{Direct}}{E_{Total}}$.

sound energy. The task is to adjust the energies of the direct sound and the reverberant part in a way that the relation between their energies is appropriate for the room situation and that the reduction of the sound level by the synthesised impulse response corresponds to the standard level difference which is used as input data.

Table 4.1 shows the relevant quantitites for 9 example rooms. It can be seen that the portion of direct sound is very small compared to the total energy of the room impulse response. Thus, for the total level of a sound signal in one of these rooms, the diffuse sound is significantly more important. To adjust the correct level of the reverberation in the simulated impulse response for the transmission from the source room to the listener in the receiving room, the following algorithm is used:

As input data for the room, the geometrical dimensions, the position of the listener, the reverberation times in one-third octaves $T(f)$, and the reverberant part of a measured binaural room impulse response $p_{rev}(t)$ are used. From this, the equivalent absorption area A is calculated by

$$A = 0.163 \frac{V}{T(500\text{ Hz})} \tag{4.5}$$

With the distance between the source (wall) and the listener $d_{S,R}$, the relation between the direct sound and diffuse field energy can be obtained by

$$\frac{E_{direct}}{E_{diffuse}} = \frac{16\pi d_{S,R}^2}{A} \tag{4.6}$$

In the next step, the absolute energy of the reverberation has to be calculated. The reverberation, normally, adds energy to the sound field depending on the reverberation times in different frequency bands. A reverberation time of 1 s, e.g., raises the sound pressure in the receiving room by 3 dB since the standardised level difference D_{nT} refers to a reverberation time of 0.5 s. The reverberant part of the impulse response, normally, has a flat spectrum at the beginning. In the different frequency bands, it, then, decreases in time corresponding to the reverberation times. The frequency bands, therefore, in total contain different amounts of energy. Here, the reverberant part of the impulse response is preprocessed in a way that the overall spectrum is white which means that to all frequency bands the same amount of energy is added. To consider the influence of the reverberation, the level difference D_j per transmission path is calculated by

$$D_j(f) = D_{nT,j}(f) - 10\log\frac{T(f)}{0.5\ \mathrm{s}} \tag{4.7}$$

where f denotes the mid frequencies of the one-third octave bands and $T(f)$ the reverberation times of the reverberant part of the measured impulse response. From this, the transmission coefficient τ is obtained by

$$\tau(f) = \frac{1}{\sum_{j=1}^{5} 10^{\frac{D_j(f)}{10}}} \tag{4.8}$$

First, the reverberant part is convolved with the time signal of the transmission coefficient τ to provide the correct spectral load of the reverberation process.

$$p_{rev,c}(t) = p_{rev}(t) * \tau(t) \tag{4.9}$$

The absolute energy of the reverberant part is, then, calculated as follows:

$$\begin{aligned} E_{p_{rev,c}(t)} &= \sum_t p_{rev,c}^2(t) \\ E_{\tau(t)} &= \sum_t \tau^2(t) \\ p_{rev,c}(t) &= p_{rev,c}(t)\cdot\sqrt{\frac{E_{\tau(t)}}{E_{p_{rev,c}(t)}}} \end{aligned} \tag{4.10}$$

I.e., the energy of the convolved reverberant part is adjusted to the same energy as the sum of the transmission coefficients τ.

Afterwards, the energy of the direct sound cluster p_{direct} is adjusted by

$$p_{direct}(t) = p_{direct}(t)\sqrt{\frac{E_{direct}}{E_{diffuse}}} \tag{4.11}$$

Figures 4.3 a - c show the single processing steps. The binaural room impulse response is preprocessed in a way that the direct sound is deleted and the initial frequency response of the remaining binaural room impulse response is flat (Figure 4.3 a).

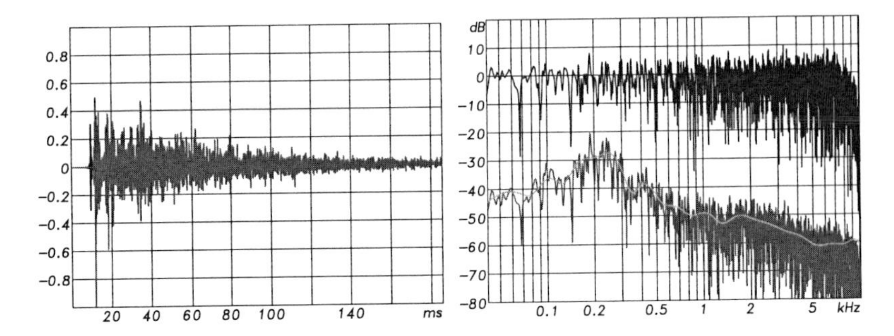

(a) Measured binaural room impulse response, direct sound and reverberation are marked in different colours

(b) Spectra: original reverberant part (blue, top) and reverberation shaped by $\sum \tau_j$ (red, bottom). $\sum \tau_j$ is displayed as green, smooth curve.

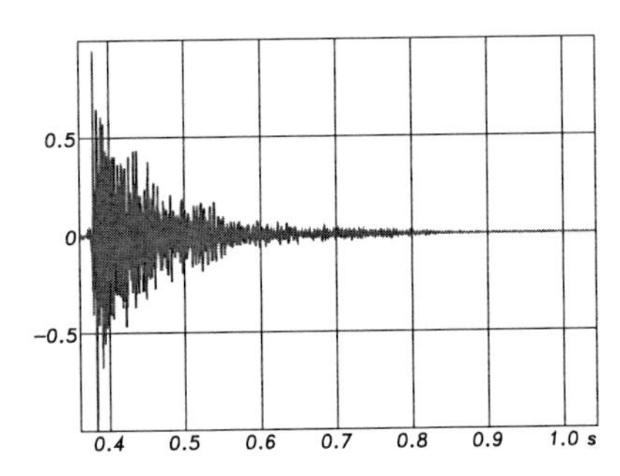

(c) Calculated impulse responses for transmission between source and receiving room for left and right ear.

Figure 4.3: Processing of the reverberant sound field

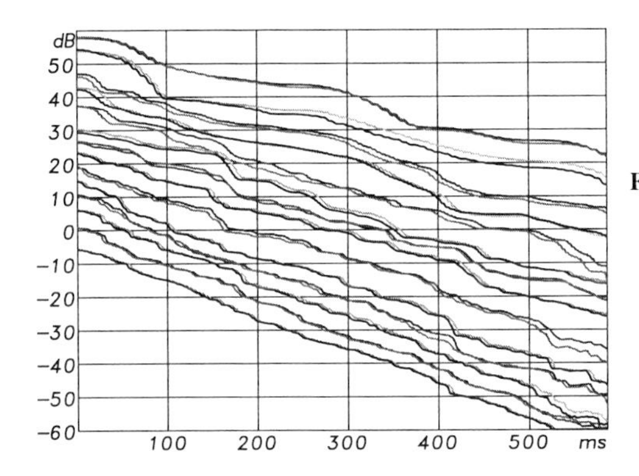

Figure 4.4: Comparison of integrated impulse responses before and after frequency shaping of the reverberant part of the impulse response. Frequencies of the displayed curves range from 50 Hz (top) to 1 kHz (bottom). The curves have been shifted in level to allow a better view.

It has to be verified that due to the frequency shaping of the reverberant part, the reverberation times are not influenced. This can be seen in Figure 4.4. The integrated impulse responses (the reverberant part) between 50 Hz and 5 kHz are displayed (pairwise) before and after convolution with the transmission coefficient. It can be seen that there are only slight changes and, thus, the reverberation is not modified.

Simplifications In this approximation, several simplifications are made. At first, the transfer functions $f_{\tau,i}$ are valid for point to point transmission only, but extended walls are present in our situation. In the receiving room, the simplification is made that the sound is apparently radiated from one point representing the whole bending wave pattern on the wall. The spectrum is exact, but the running wave pattern on the wall is replaced by a point source at the centre of the wall with a linear phase. It could be shown in [Hol02] that a more exact model using up to 81 point sources on a wall did not bring significant improvements.

For the receiving room, a measured impulse response is used instead of a room acoustical simulation. Thus, the geometries and absorption might not totally fit to the properties desired by the user of the auralisation software. However, in this application, a selection out of 9 different rooms can be made which reduces the error.

The transfer functions $f_{rev,j}$ representing the room transfer function between 'the wall' and the receiver are assigned to the same point of radiation instead of using different room impulse responses for each source receiver combination (radiating point source and listener). The assumption is made that the reverberant part is relatively constant for different positions.

As input signals for the auralisation, pre-recorded signals are used which already contain the room acoustical properties of the recording room. For example, a recording of dance music from a stereo set already contains the room acoustical properties of the actual room in which it was recorded. Thus, it is not possible, to select the source room and the ‘dry’ source independently. This, however, is easily to be improved by carrying out a convolution of a dry recorded source signal with a binaural room impulse response as is used for the receiving room. Here, it was decided to leave out this convolution because of less consumption of calculation time, less effort when doing the recordings, and simplicity and transparency of the software for the user. In the end, it makes no difference where the reverberation is introduced in the linear signal flow. After all, the level, colouration, and reverberation is exactly calculated.

Chapter 5

Impact sound insulation

This chapter describes the algorithm and creation of filters for the auralisation of impact sound insulation. For an exact calculation of the sound field at the ears of a listener in the receiving room, a detailed model has to be developed which includes quantities as the force-time signal of the source, the impedance of source and floor, the radiation efficiency, characteristics of the receiving room, etc. A possible detailed model is displayed in Figure 5.1. Quantities as radiation efficiencies or floor impedances, however, are normally not given for building products. Commonly, only the normalised impact sound pressure level L_n and the structure-borne reverberation time are given. It is, thus, desirable to base an algorithm for auralisation on these commonly available quantities. To illustrate this, the model in Figure 5.2 should be looked at. A source (tapping machine, walker, etc.) is considered as an ideal force source (F_0) acting upon an impedance (two-port $Z(f)$) which is given by the mechanical impedance of, e.g., leg and shoe of a walker or the impedance of one hammer of a tapping machine. These elements can be transformed into a real force source with a force F_0' and an inner impedance $Z_s(f)$. This real force source is connected to a two-port with the input quantities F_f and v_f (see Figure 5.3). The quantities at the output are the sound pressure p and particle velocity v in the receiving room. For measurements, say, of a heavy-weight concrete floor with a standard tapping machine, the inner impedance of the source Z_s can be neglected, since it is much smaller than the input impedance of the two-port which is formed by the floor impedance. Thus, this model can be taken as a description of the measurement of the impact sound pressure level L_n. It can be said that

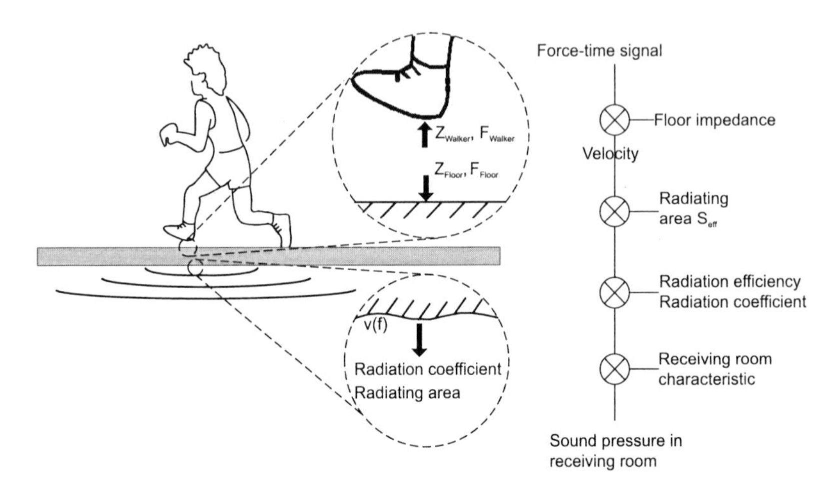

Figure 5.1: Detailed model for the auralisation of impact sound insulation

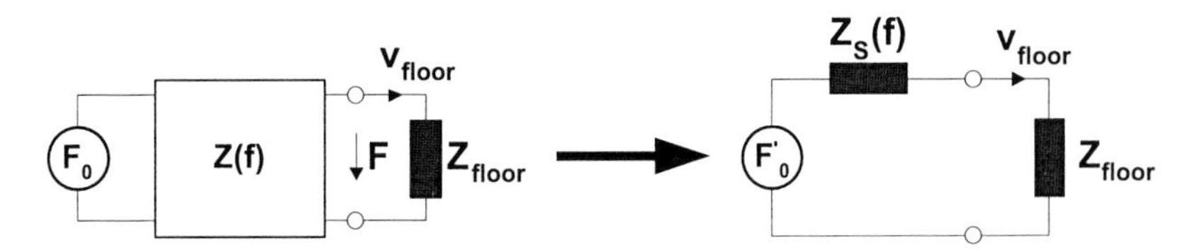

Figure 5.2: Model of an ideal force source with an inner impedance acting upon the floor impedance and the transformation into a real force source

quantities describing the excitation of the floor by the source, the radiation in the receiving room, and the receiving room characteristics are already contained in the two-port and, thus, in the impact sound pressure level L_n for the given configuration in the measurement. The task is to remove the influence of the source (tapping machine) and the receiving room on the quantity L_n and replace it by the desired force source and receiving room.

The principle for auralisation of impact sound insulation is, thus, basically the same as for auralisation of airborne sound insulation except for the excitation mechanism in the source room. In short, the sound pressure in the receiving room can be obtained by

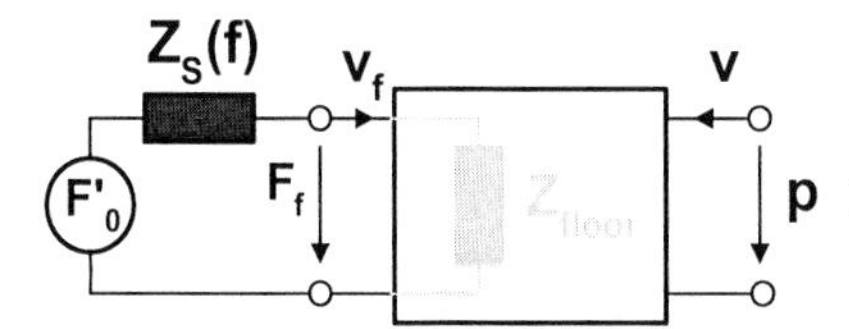

Figure 5.3: Model of a real force source connected to a two-port representing the floor construction and the receiving room

1. dividing the normalised impact sound level by the force spectrum of the tapping machine
2. multiplying the result with the complex force spectrum of the chosen force source (e.g. a walking person)
3. modelling the sound field in the receiving room (reverberation etc.) in an appropriate way

The modelling of the sound field in the receiving room is just the same as described in Chapter 4.

Another important issue is the feedback between the source and the structure in the source room. Whereas for airborne sound the feedback between source and load (the radiation impedance) can be neglected due to the impedance relation ($Z_s << Z_{floor}$), for impact sound excitation, the impedance relation may be smaller, especially for excitation of light-weight structures by a tapping machine. Investigations carried out in the past have shown that measurements of light-weight and heavy-weight floors with a tapping machine are not comparable (see e.g. [SM99]). It is, thus, necessary to measure the impedance of the sources and floors to be auralised. The set-up and the method for measuring forces and impedances are also described in the following. Besides, literature can be found in [Tha01] and [Tha04].

A first simple model In a first step, a simple auralisation without considering the dynamic interaction between source and floor is developed. A model for the sound pressure level in the receiving room from a tapping machine (TM) in the source room can be taken from [Ger90] as

$$p_{TM}^2 = \frac{4\rho^2 c^2 T_s \sigma}{2\pi 2.2 m A Z_s} F_{TM}^2 \quad (5.1)$$

with p_{TM} denoting the sound pressure in the receiving room caused by a tapping machine, σ the radiation efficiency, m the mass, A the area, F_{TM} the force injected by the tapping machine, and T_s and Z_s the reverberation time and impedance of the structure (floor), respectively. One can see that the relation between sound pressure p_{TM} and force F_{TM} is determined by a term which is only dependent on the floor construction. Assuming a linear behaviour, the sound pressure resulting from a different force source (FS) can, thus, be obtained by

$$p_{FS}^2 = \frac{p_{TM}^2}{F_{TM}^2} F_{FS}^2 \quad (5.2)$$

with p_{FS} and F_{FS} denoting sound pressure and force caused by an arbitrary force source. One can find an expression for the sound pressure in the receiving room depending on the normalised impact sound level from different transmission paths and the reverberation time considering equation (5.2) which yields

$$p_{FS}^2 = \frac{F_{FS}^2}{F_{TM}^2} \sum_{j=1}^{5} 10^{\frac{L_{n,j}}{10}} p_0^2 \frac{T}{T_0} \frac{V}{30} \quad (5.3)$$

where $L_{n,j}$ are the normalised impact sound levels of the independent paths and p_0 is the reference sound pressure.

The quantities to be obtained for the auralisation are the normalised impact sound pressure levels $L_{n,j}$ for the different transmission paths, the force of the tapping machine F_{TM} and the desired force source to be auralised F_{FS}, and the receiving room reverberation time T. The impact sound pressure levels are obtained by simulation software, the forces are measured. For here, the results shall be used in advance to the description of the measurement set-up and method which follows later. Figure 5.4 shows the one-third octave spectra of the measured forces of the tapping machine, the modified tapping machine, and a rubber ball according to prEN ISO 140-11 [ISO03]. The spectra of the tapping machine and the modified tapping machine are equal at low frequencies. Then, above 125Hz the force of the modified tapping machine gets significantly smaller due to the rubber layer between

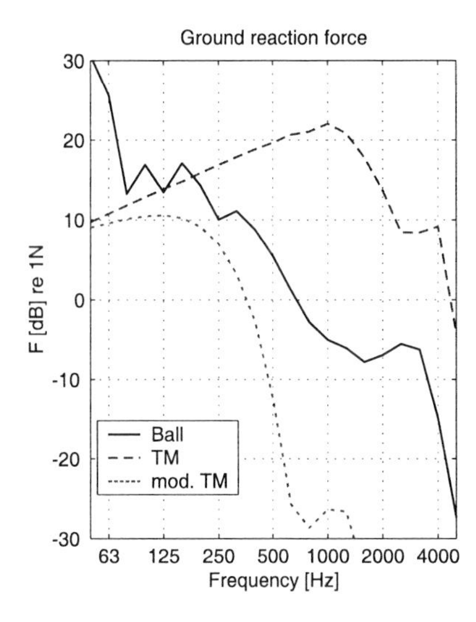

Figure 5.4: Ground reaction forces of a tapping machine, a modified tapping machine and a rubber ball in one-third octave bands

hammer and floor. It should be remarked that due to properties of the measurement setup, the measured force of the tapping machine seems to be too low above 1kHz. For use in auralisation of common sources, this is not a problem since this frequency range can be neglected.

For the auralisation, a force-time signal of the necessary length is constructed from several measured force pulses of the source. As an example, from two or three force pulses of one hammer of a tapping machine, a force-time signal is constructed by appending the pulses at an interval of 0.1 seconds. To make the signal sound more naturally, jitter in time and amplitude can be introduced. In a first try, an auralisation of the four room situations shown in Figure 5.5 was carried out. The situations were modelled in a building acoustics software for calculating sound insulation of room situations from building products [Dat99] which delivers the normalised impact sound levels to be used for the auralisation. The impact sound levels of the auralised signals were calculated from the auralised audio files by an evaluation software (Artemis, Head acoustics). They were, then, compared to the normalised impact sound levels obtained by the modelling software which were used as input parameters for the auralisation. At first, the levels of the auralised signals are not absolutely calibrated. Only the differences between the room situations are considered. For a better comparison, the measured levels are referenced to the room situation with the highest weighted impact sound level $L_{n,w}$. Table 5.1 shows a comparison between $L_{n,w}$, $L_{n,w+Ci}$, and the levels of the auralised signals for the tapping machine (TM) and the modified tapping machine (mod.TM).

Input parameters		Auralisation results		
Floor/Covering	$L_{n,w}$	$L_{n,w} + C_i$	L_{TM}	$L_{mod.TM}$
Chipboard	52dB	88dB	58dB	54dB
Cement	60dB	65dB	64dB	55dB
Concrete	76dB	57dB	75dB	58dB
Aer. concrete	99dB	53dB	99dB	76dB

Table 5.1: Input quantities and impact sound levels from auralisation

It can be seen that the values for $L_{n,w}$ and the auralised level of the tapping machine correspond quite well for bare floors, but not similarly well for the floors with additional layers. The modified tapping machine gives rather different level results which correspond better with $L_{n,w+Ci}$. This can be explained by the forces of the two sources. Whereas the tapping machine produces a rather broad force spectrum, the modified tapping machine only contributes energy up to, say, 400 Hz. Since $L_{n,w+Ci}$ focuses more on lower frequencies, its result seems to be more reasonable than $L_{n,w}$.

The sound examples described above can be found under [Tha] The next step would be to consider the dynamical behaviour of source and structure. For it, characterisations of sources by forces and impedances must be made. In the next chapter, different approaches are presented.

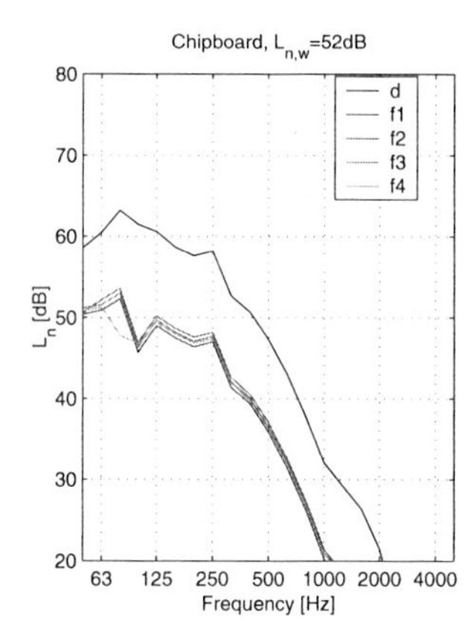

(a) Concrete, additional layer, $L_{n,w} = 52 : \mathrm{dB}$

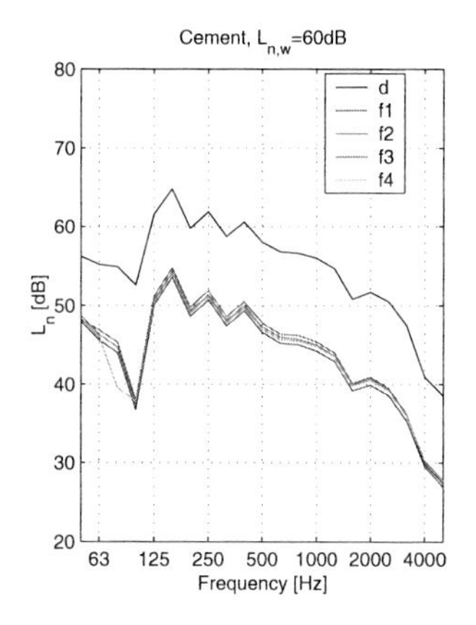

(b) Concrete, additional layer, $L_{n,w} = 60 : \mathrm{dB}$

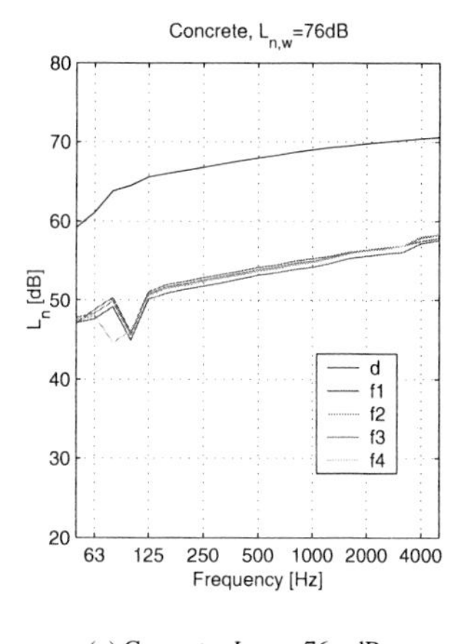

(c) Concrete, $L_{n,w} = 76 : \mathrm{dB}$

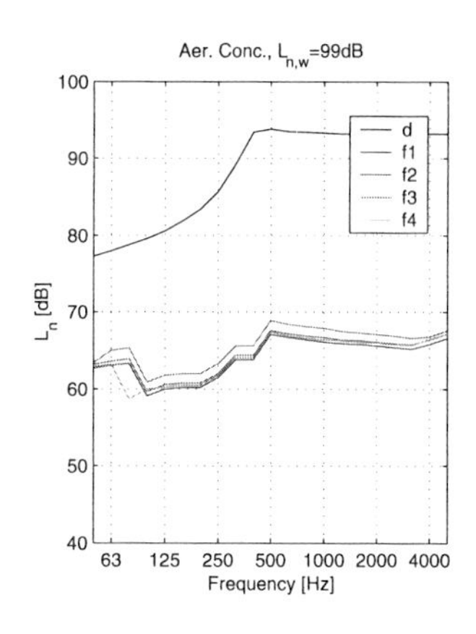

(d) Aerated concrete, $L_{n,w} = 99 : \mathrm{dB}$

Figure 5.5: Normalised impact sound levels for 4 room situations

5.1 Measurement Methods and Setup

This section describes measurement methods and setups used for determination of impact forces and source impedances. In a first approach, forces were measured using an aluminium slab with force transducers underneath as shown in Figure 5.6. An aluminium slab

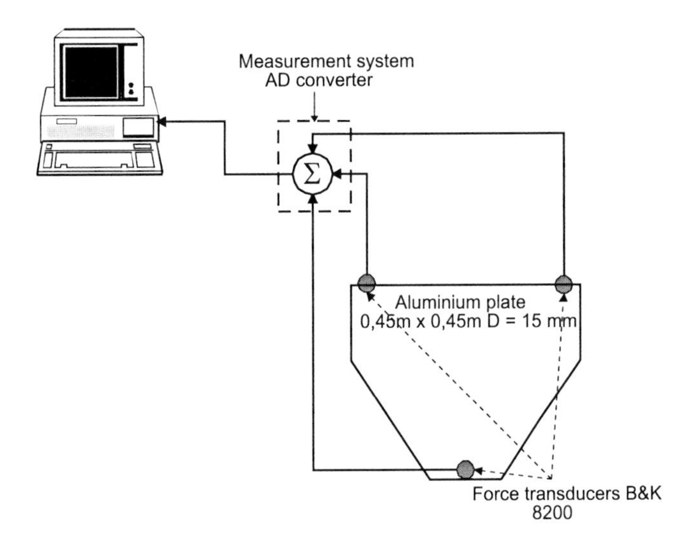

Figure 5.6: Setup for measuring ground reaction forces. The forces of the three transducers are summed up.

of the dimensions 0.45 m x 0.45 m, shaped slightly triangular with three force transducers underneath is mounted on a thick concrete floor. The floor beneath can be considered as infinitely high impedance. The signals of the three force transducers are summed and represent the force which acts on the slab. Figure 5.7 shows force time signals measured with this setup for a walking and a running person. The differences between walking and running can be seen clearly: For walking, the force time signal is longer and, thus, contains less high frequency content than the signal for running which has a higher first impact. The

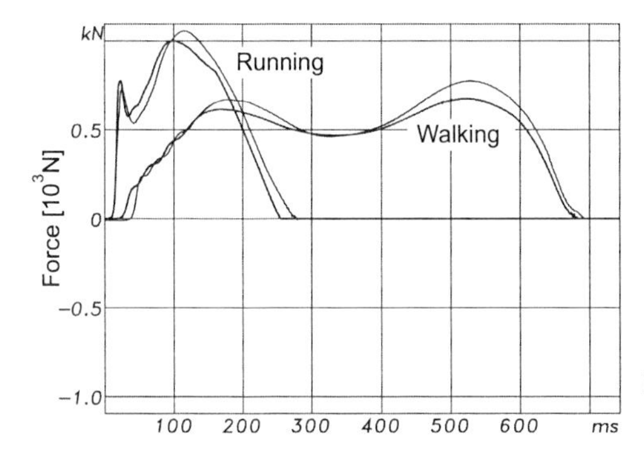

Figure 5.7: Ground reaction forces of a walking and a running person.

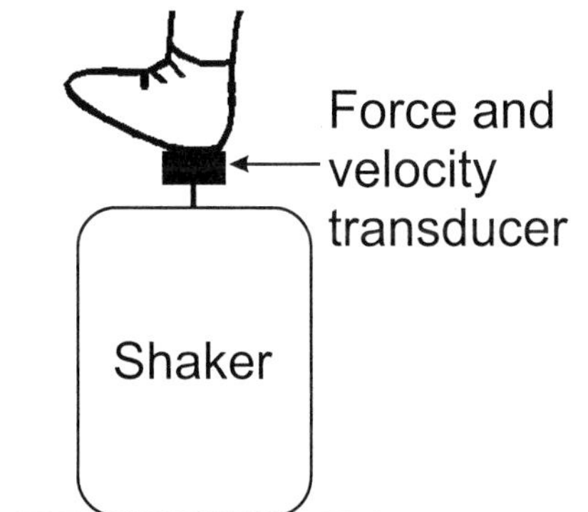

Figure 5.8: Measurement of the impedance of a walker in a static condition. The foot is excited by a shaker and the force and velocity are determined and devided.

measurement setup, however, has one drawback: Due to eigenmodes of the slab, the measurement result is distorted. The eigenmodes for the used slab start at about 150 Hz. As a solution, the stiffness of the slab must be increased (different material) or the size must be reduced to increase the frequency of the eigenmodes and shift it in a frequency range which can be neglected.

To measure the impedance, the setup as shown in Figure 5.8 was used. The foot is placed upon a small aluminium slab with a force and velocity transducer mounted underneath. The force and velocity are measured while the foot is excited by a shaker and divided to obtain the mechanical impedance of the foot. Figure 5.9 shows the results of measurements on a walker with shoes and barefoot. For the barefoot walker, it can be seen that there is a stiffness controlled region (up to 200 Hz) followed by a mass controlled region. Below 50 Hz, the impedance also seems to be dominated by a mass. For the walker with shoes, the mass controlled region seems to disappear. Between the measurements, the pressure of the foot on the measurement setup was changed (indicated by arrows in Figure 5.9). It can be seen clearly that the stiffness is changed due to the compression of the flesh/rubber underneath the foot. This, of cause, must be considered since for a walking person the impedance changes during one footstep which introduces nonlinearities. It has to be evaluated if these nonlinearities can be neglected and the impedance may be characterised by one curve only.

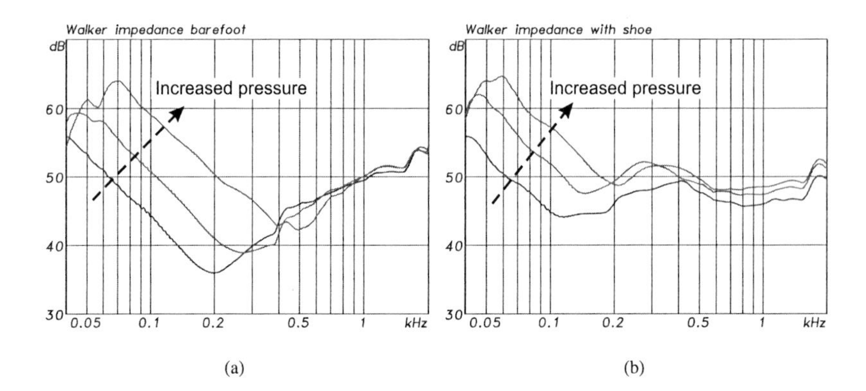

Figure 5.9: Impedances of a walker barefoot (a) and with shoes (b). The 3 curves represent the impedances for different pressures acting upon the measurement system. With higher pressures, the stiffness of the foot (flesh/rubber is compressed) increases.

To better fit the real conditions, it was decided to measure the impedance in a dynamic condition, i.e., during walking. For it, the following considerations were made: The source is considered as a real force source as shown in Figure 5.2. Assuming a linear behaviour, this real source can be described by its open-circuit force and short-circuit velocity analogue to a real voltage source as shown in Figure 5.10. For one frequency, this characteris-

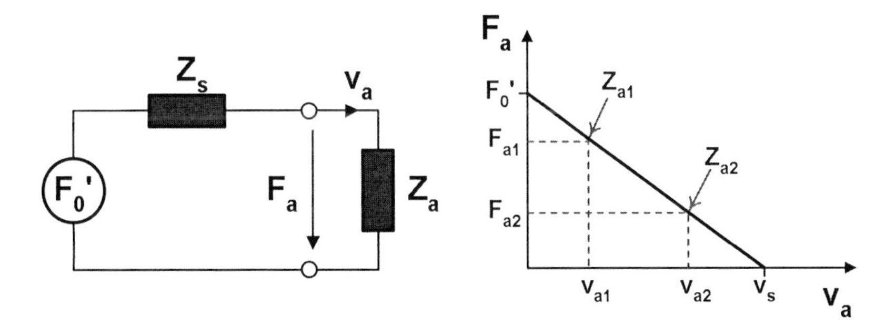

Figure 5.10: A real force source and its characteristic derived from two measurements with load impedances Z_{a1} and Z_{a2}

tic describes the relation between the force and the velocity for a certain load impedance. The inner impedance Z_s and the open-circuit force F'_0 can be determined from the force and velocity measured at two different load impedances. For a frequency dependent source and impedance, a characteristic as in Figure 5.10 has to be derived for each frequency. The real force source behaves as follows: for floors with low impedances, a high velocity is introduced with already a small force. The higher the floor impedance, the lower the velocity and the higher the force. The power delivered by the source reaches a maximum if the load impedance and the inner impedance are equal. For a floor with an infinitely high impedance (like a concrete floor on the foundation of a building) the velocity is zero and the force is F'_0. The velocity reaches its maximum v_s at $Z_a = 0$. From two measurements of the force F_{a1}, F_{a2} and the velocity v_{a1}, v_{a2} over two load impedances Z_{a1} and Z_{a2} connected to the real force source the ideal source F'_0 and the impedance Z_w can be calculated as follows:

$$Z_s = \frac{F'_0}{v_s} = \frac{F_{a1} - F_{a2}}{v_{a1} - v_{a2}} \tag{5.4}$$

$$F'_0 = \frac{Z_{a1} + Z_s}{Z_{a1}} F_{a1} = \frac{Z_{a2} + Z_s}{Z_{a2}} F_{a2} \tag{5.5}$$

A measurement setup was to be developed which allows the measurement of the force and the velocity 'below' a human walker connected to 2 different load impedances. One important condition to be fulfilled is to apply identical excitation forces for the measurements on the 2 different impedances. This means that the person under test has to walk on the measurement slab twice in exactly the same manner, which is nearly impossible. As an approximate solution, several measurements must be carried out and mean values for the impedance and the open-circuit force must be derived. Additionally there is a problem with the signal to noise ratio during the measurement. The signals to be measured are the force and the velocity induced by the walker. Especially for higher frequencies the force and the velocity are at a very low level. To obtain reasonable signal-to-noise ratios, the connected load impedance should be in the same dimension as the inner impedance of the walker (matched impedances to produce high output signal). Here, the load impedances are realised by a rubber and a cork layer underneath the piezo elements. The frequency dependent impedances can be seen in Figure 5.15 c. To make first experiences with this measurement method, a model of the walker and the measurement set-up was built out of small springs and masses (see Figure 5.11).

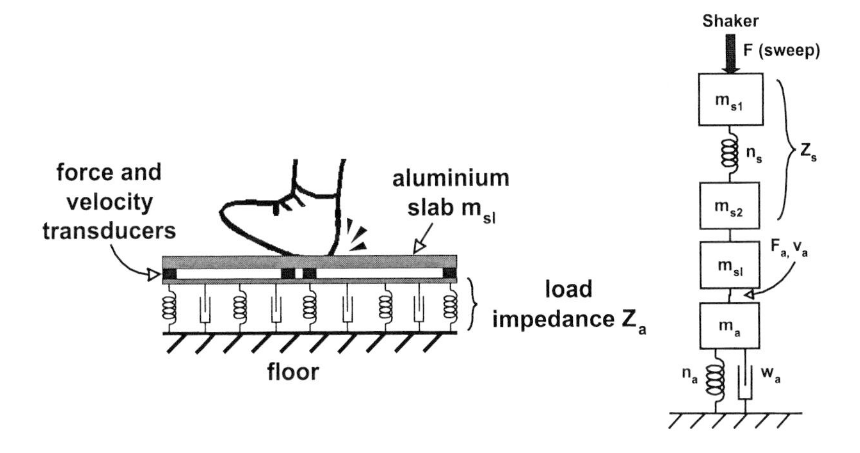

Figure 5.11: Measurement setup for determination of impedance and open-circuit force of a source and a model for a human walker which was used for experimentation.

The walker was modelled as mass-spring-mass system with $m_{s1} = 400$ g, $m_{w2} = 20$ g and $n_w = 3.4 \cdot 10^{-6}$. The model was taken from an article published by Scholl [SWSK02]. The elements differ from that model just for practical reasons (available material etc.). The aluminium slab which is to be used in the real measurement on a human walker is modelled by a steel cylinder with a mass of $m_{sl} = 380$ g. The load impedances consists of a mass $m_{a1} = 590$ g resp. $m_{a2} = 40$ g and a spring with a compliance of $n_{a1} = 2.32 \cdot 10^{-7}$ m/N resp. $n_{a1} = 3.3 \cdot 10^{-7}$m/N. As excitation source a shaker was mounted on top of this model making sure that for each two measurements the model was excited by identical forces. The excitation signal was shaped in a way, that the problems with low signal-to-noise ratios at higher frequencies are reduced.

Figure 5.12 a-c show the results of the force (upper curves) and the velocity (lower curves) measurement at two different load impedances, the calculated impedance, and the open-circuit force of the (modelled) walker according to Equation (5.4) and (5.5).

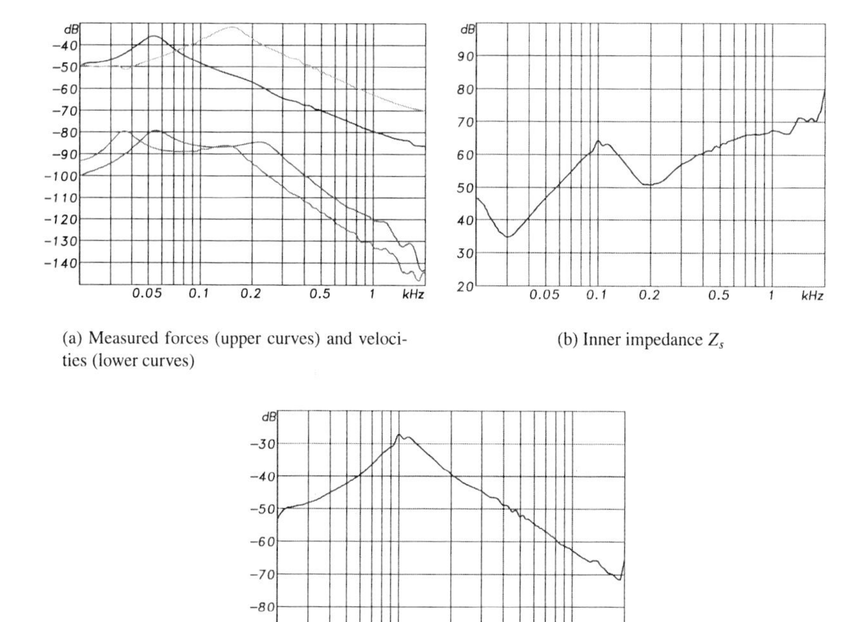

(a) Measured forces (upper curves) and velocities (lower curves)

(b) Inner impedance Z_s

(c) Open-circuit force F'_0

Figure 5.12: Results of the model measurements for the force and velocity over 2 load impedances (a), the calculated inner source impedance (b) and the open-circuit force (c)

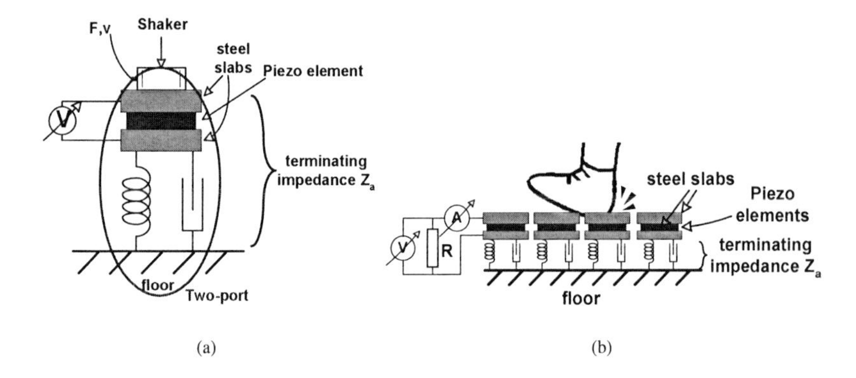

Figure 5.13: Setup for calibration of 1 piezo element (a) and setup for determination of source impedance in dynamic condition (b).

Still, the problem to be solved was the eigenmodes of the slab for the measurement. The slab has to be small enough to have its resonance frequencies above the interesting frequency range (about 1 kHz) and big enough for a person to step on. Since these criteria were hardly to be fulfilled, the slab was divided into several small slabs with a size of 10cm in square. Each of it had to have force and velocity transducers underneath and the resulting signals for each slab were to be added. Since a multi-channel measurement equipment and a high amount of transducers would have been needed, it was tried to use piezo elements for measurements of the force and the velocity. A measurement setup in total and the calibration of one element is shown in Figure 5.13. The piezo element together with the terminating impedance and the floor is considered as two-port. A calibration of the elements with known terminating impedances and measurements of the voltage delivers the desired quantities F_a and v_a.

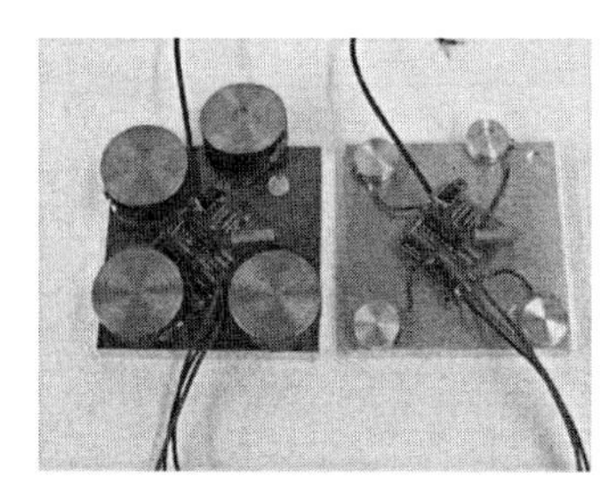

(a) Piezo setup for 1 slab of 10 cm in square

(b) Calibration of a piezo element using a shaker

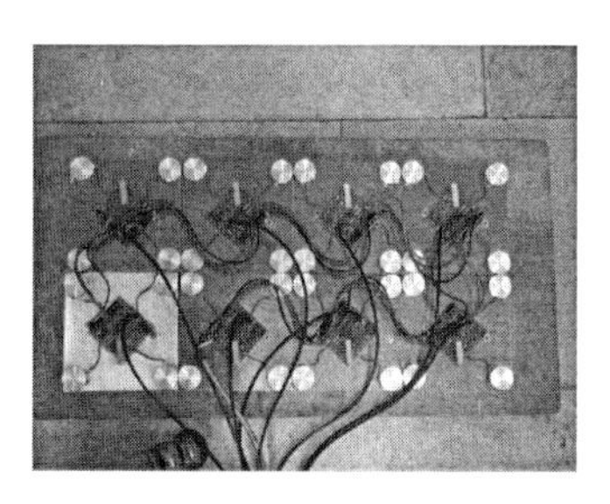

(c) Setup of one force-impedance measurement element for 1 load impedance

(d) Total measurement setup using 2 different impedances and a gangway

Figure 5.14: Setup using multiple piezo elements

Since this method did in the end not fulfil all requirements, the detailed description of calibration and measurements is left out here and can be found in [TB03] and [Bra03]. Figure 5.14 a-d show the setup. For one element, 4 piezo elements were used and the signals are added by a small electronic circuit. In Figure 5.14 b, the setup for determining the two-port parameters can be seen. While excited by a shaker, the force and velocity on top and the voltage over the piezo element is measured. 8 of these elements form a measurement setup for one load impedance which is big enough for a person to step upon and the single elements are small enough to shift their resonance frequencies beyond 1.5 kHz. To verify the setup, the shaker was used as source. A synthesised signal (sine-shaped pulse) simulating a single footstep was 'replayed' by the shaker. Force and velocity on top of the slab were measured by transducers and in parallel calculated from the voltages measured over the piezo element. From these quantities, the source impedance and open-circuit force were calculated. For a correct verification, the quantities Z_s and F'_0 derived from the force and velocity measurement and from voltage measurement and calculation should be identical. Figure 5.15 a-e shows the results. Note that in Figure 5.15 e, the open-circuit force is calculated from the forces F_{a1} and F_{a2} measured over both load impedances. Normally, calculation from one of these quantities would be sufficient (see Equation (5.5) but a calculation from both quantities with identical results gives an additional verification.

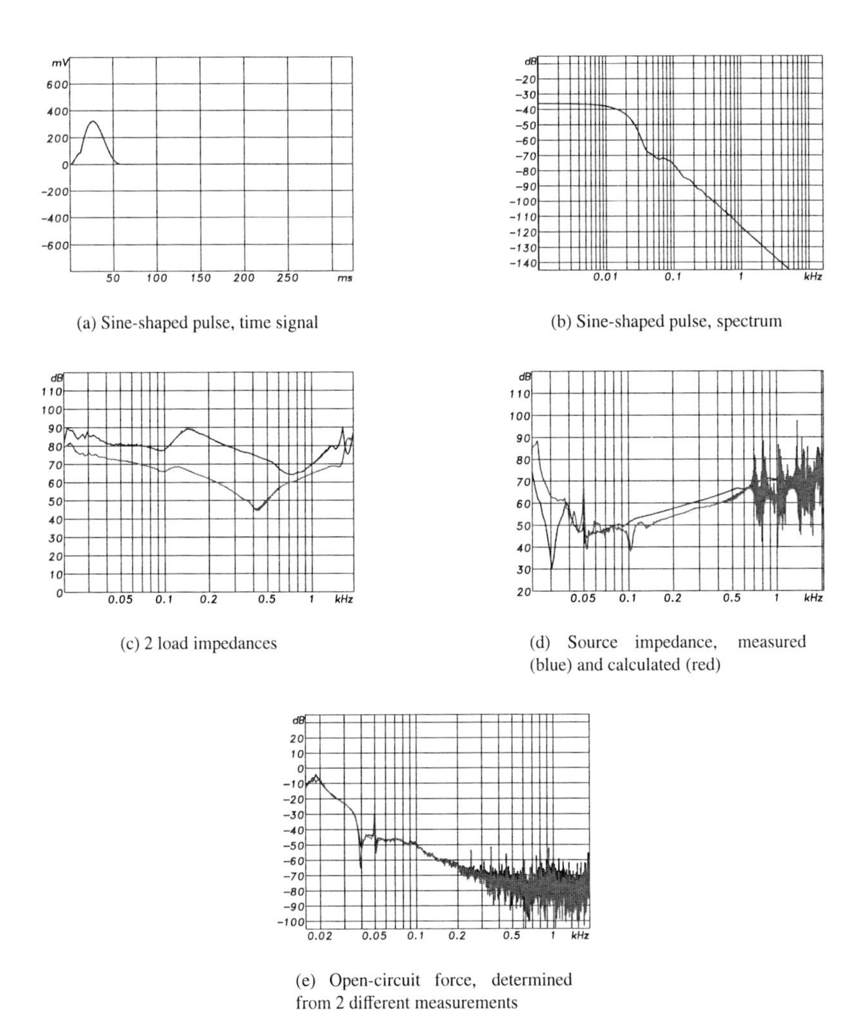

(a) Sine-shaped pulse, time signal

(b) Sine-shaped pulse, spectrum

(c) 2 load impedances

(d) Source impedance, measured (blue) and calculated (red)

(e) Open-circuit force, determined from 2 different measurements

Figure 5.15: Verification of piezo measurement setup. Shaker acts as source and emits sine-shaped pulse as seen in a and b. c shows the load impedances connected to the source. d compares the source impedance determined by vibroacoustic measurement (correct) and calculation from piezo elements (to be verified). e shows the calculated open-circuit force calculated from both forces measured over the load impedances. Ideally they should be identical.

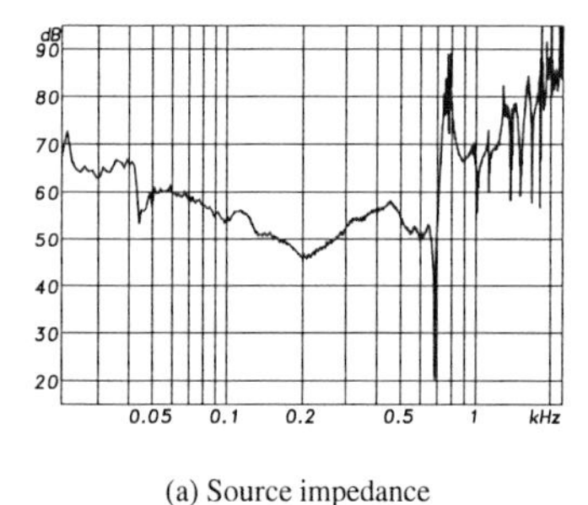

(a) Source impedance

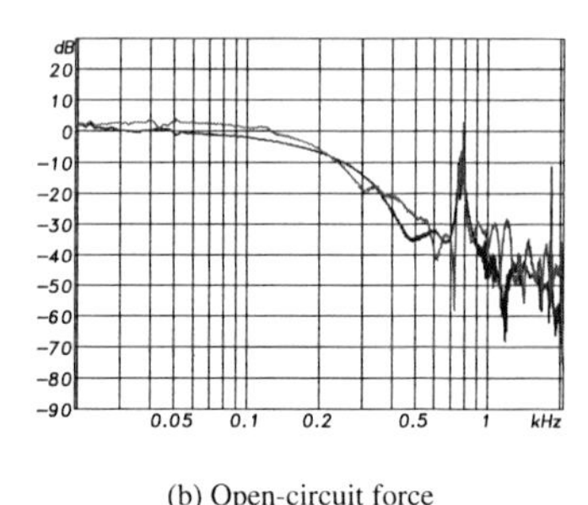

(b) Open-circuit force

Figure 5.16: Measurement of the modified tapping machine. a shows the source impedance, b the open-circuit force calculated twice from both forces over the load impedances. (Should ideally be identical)

Since the method worked in principle, as next step, a real source was to be measured. It was decided to use the modified tapping machine according to [ISO03]. It consists of a standard tapping machine with a rubber material of defined stiffness laid underneath. The tapping machine was placed on top of a measurement element and the voltage of the piezo elements were recorded. From this, the force and source impedance of the modified tapping machine was calculated. The results can be seen in Figure 5.16. The results seem to be reasonable, as the shape of the impedance curve fits to the results from Scholl. The open-circuit force decreases at frequencies around 200 Hz due to the rubber layer underneath the tapping machine and fits to the result from a different measurement method shown in Figure 5.4. At frequencies around 700 Hz, however, strong distortions are introduced the source of which is still not clear. During several other measurements, drawbacks of the setup, mainly due to mechanical problems, were found. The piezo elements broke when sources with high forces (standard tapping machine) were measured. Also the dynamical range at high frequencies was not sufficiently high, thus, noise was introduced.

Thus, it was thought about using different materials for the slab to step on which should be stiff and light-weight (high resonance frequencies). Honeycomb sandwich panels, a composite material from 2 thin aluminium or Carbon fibre slabs with honeycomb-structured aluminium in-between were used to measure the forces of the tapping machine, the modified tapping machine and a rubber ball. The results have been shown previously in this work and can be seen in Figure 5.4.

First experiments show that the material is useful for this task. A measurement setup consists of one slab with 3 force and acceleration transducers underneath and again 2 different load impedances.

With this promising approach, it is possible to realise the source characterisation. Thus, different sources like human walkers, falling toys, etc., can be recorded and auralised in the future. This work still has to be done.

Chapter 6

Verification

In this chapter, a detailed verification of the algorithm for auralisation of airborne sound is presented. First of all, the most important parameter - the level - is verified to be reproduced correctly. On the basis of a real building situation - two office rooms - a verification according to EN ISO 140-4 [ISO98] was carried out. Additionally, a listening test with auralised sounds and sounds recorded in the real situation was done to verify the correct reproduction of the colouration and the naturalness of the auralised sounds. In parallel, psychoacoustic parameters were calculated.

6.1 Level Reproduction

The setup for the measurements was as follows: The sounds which would be heard in the source and the receiving room were replayed by the auralisation software and directly fed from the PC soundcard into a sound level analyzer (Norsonic 110). As source room signal, a white noise signal was used. The level in one-third octave bands of both signals was measured and the level difference $D = L_{source} - L_{aura}$ was calculated. With the reverberation times of the receiving room which were derived from the room impulse response, the standardised level difference $D_{nT,aura} = D + 10 \log \frac{T}{0.5\,\mathrm{s}}$ was determined. A comparison of the input quantity $D_{nT,in}$ which was delivered by the sound insulation calculation software [Dat99] with the measurements of $D_{nT,aura}$ is shown in Figure 6.1 a. Figure 6.1 b shows the difference $D_{nT,aura} - D_{nT,in}$. It can be seen that at lower frequencies, the deviation between

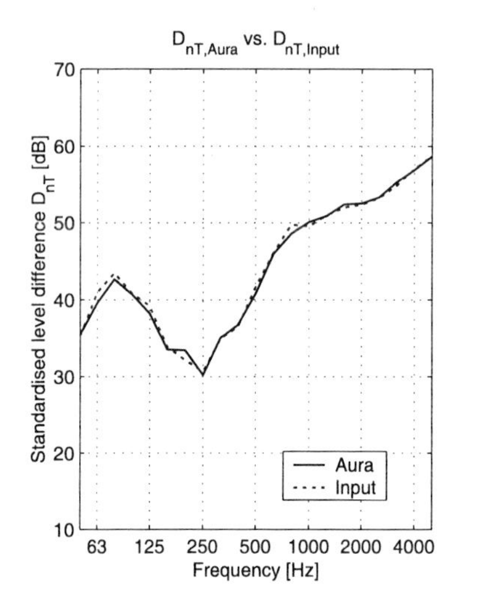

(a) Numerical input data (blue, dashed) and auralisation (black, solid)

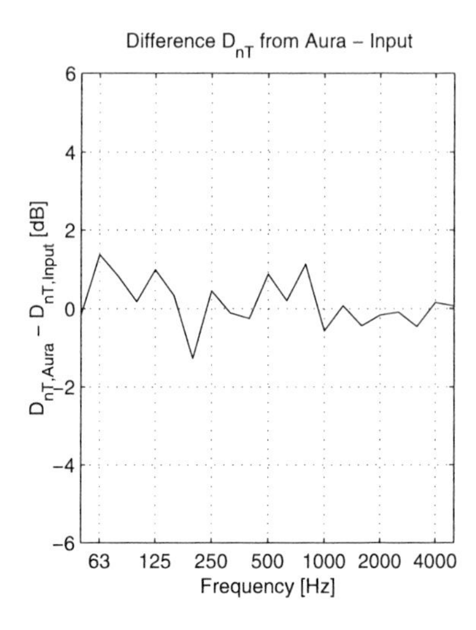

(b) Differences between input and auralisation

Figure 6.1: Measurement of the level from auralisation compared to the data used as input for auralisation. Differences at low frequencies are mainly due to effects of modes in the measured receiving room impulse response.

input and auralisation is somewhat higher. It has to be kept in mind that the auralisation is only done for one position of source and receiver. According to EN ISO 140-4, 5 positions in the receiving room have to be evaluated to compensate for the low modal density at lower frequencies.

6.2 Comparison with Real Situation

For a more correct verification, recordings and measurements on a real building situation were used. The situation consisted of two office rooms at the Institute of Technical

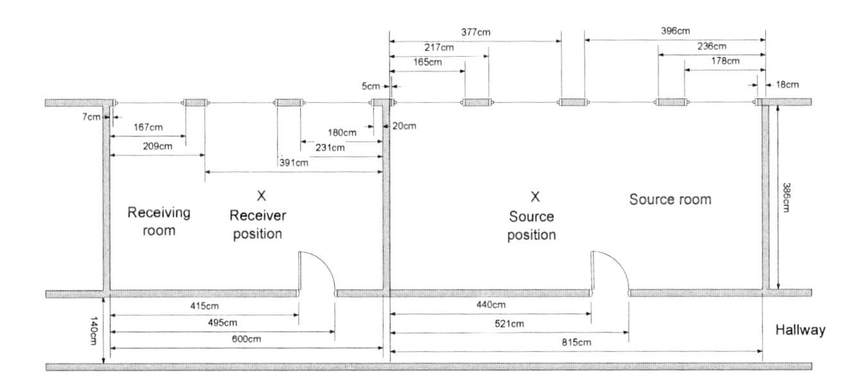

Figure 6.2: Floor plan of the building situation which was used for verification of the auralisation

Acoustics with the dimensions (L x W x H) of 8.15 m x 3.83 m x 3.25 m (source room) and 6 m x 3.83 m x 3.25 m (receiving room). Figure 6.2 shows a floor plan of the two rooms. In Figure 6.3, the interior of the rooms can be seen. The reverberation times and the sound reduction index were measured and used as input for an auralisation. Additionally, sounds of different sources were recorded with 2 dummy heads and a microphone in the source and receiving room and compared in a listening test to the auralised versions of the recordings. The reverberation times at 500 Hz are $T_{source}(500 \text{ Hz}) = 0.54$ s and $T_{recv}(500 \text{ Hz}) = 0.69$ s. For the verification, two different cases were measured: the two room situation with doors closed and the same situation with open doors in both rooms. For the latter, the main sound transmission occurs via the outer path over the hallway. In this case, the transmission via the direct flanking path can be neglected. The hallway was covered with absorptive material to avoid too much reverberation.

6.2.1 Measurements

The measurements of level differences were carried out according to EN ISO 140-4 [ISO98]. 5 microphone positions in each room and 2 speaker positions in the source room

(a) Source room (b) Receiving room (c) Hallway between offices

Figure 6.3: Pictures of the rooms used for verification of the auralisation

were measured to avoid deviations due to the low modal density at lower frequencies. The measurement setup can be seen in Figure 6.4. As source, a 2-way speaker system was used as shown in Figure 6.6 b. Sweep signals with a duration of ca. 24 s were used to measure the sound insulation. The generation of the signals was done according to an algorithm published by Müller and Massarani [MM01]. This method allows to adjust the frequency growth rate of a sweep continuously in a way that the measurement system (speaker, amplifier and AD/DA circuits) is equalised. The aim is to acquire room impulse responses which are free of colouration and achieve an almost frequency-independent SNR. The deconvolution process which transforms the sweep responses into room impulse responses is able to completely reject all harmonic distortion components. For the measurement of level differences, a sweep signal was replayed from the 2-way speaker system and the sound pressure levels in both rooms were recorded. The measured level differences $D = L_{source} - L_{recv}$ for the 2 situations (open and closed doors) are shown in Figure 6.5. The reverberation times are derived from the room impulse responses which were measured by placing the speaker system and the microphone/dummy head in the room and measuring the impulse response for this source receiver combination. In Figure 6.6 a, the reverberation times in one-third octave bands can be seen. It has to be noted that not only the position of the microphone/dummy head is relevant but also a different position of the speaker influences the result especially at low frequencies where the modal density is low. If, e.g., the speaker is positioned in the node of a room mode where the sound pressure is minimal and the particle velocity is at its maximum, the speaker 'sees' a lower radiation resistance and radiates less power for this frequency.

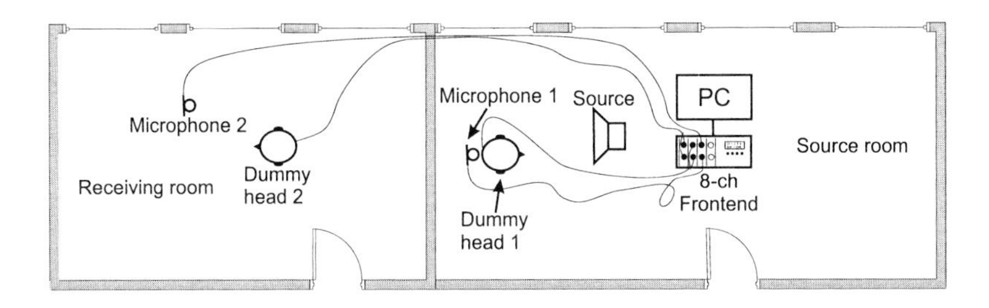

Figure 6.4: Plan of the building situation and equipment used for measurements and recordings

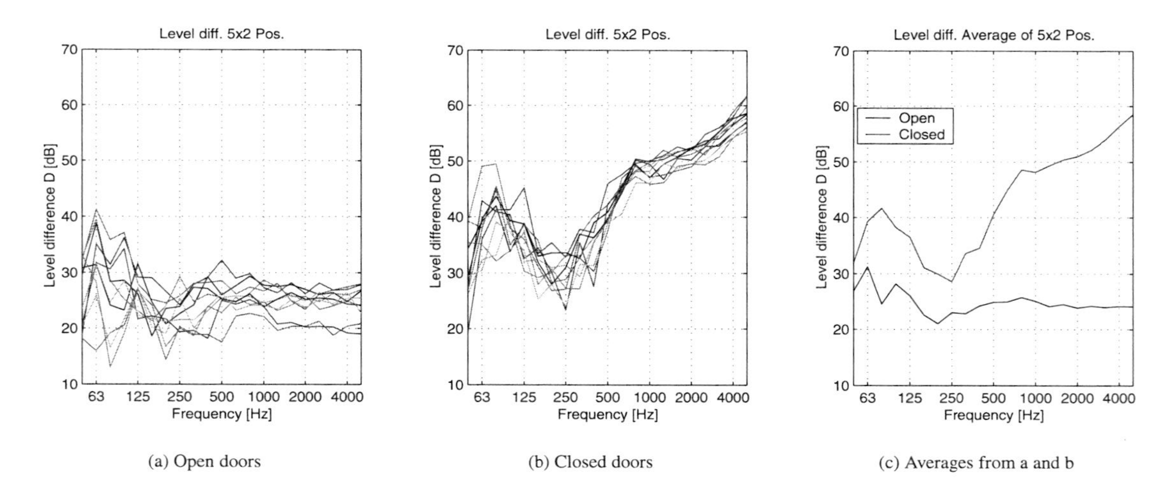

(a) Open doors

(b) Closed doors

(c) Averages from a and b

Figure 6.5: Measured level differences in a 2-room situation with 2 speaker and 5 microphone positions. Figure c shows the averages of the 10 measurements in Figure a and b.

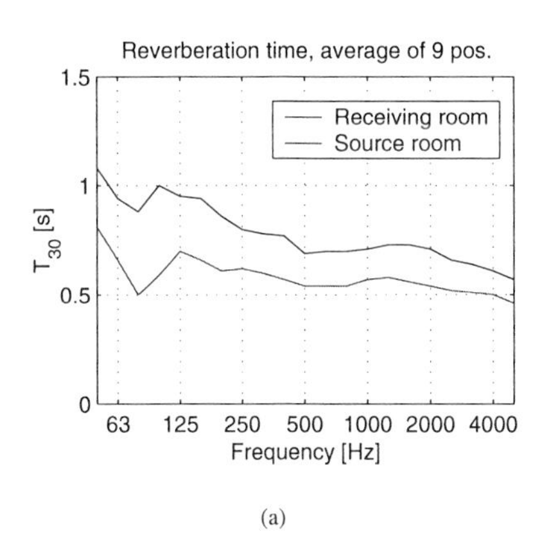

(a)

(b)

Figure 6.6: Figure a shows reverberation times (T_{30}) of the source and receiving room from 9 measurements at different positions. In Figure b, the 2-way speaker system used for measurements can be seen.

6.2.2 Recordings and Auralisation

For the auralisation, the measured standardised level differences D_{nT} and reverberation times in one-third octave bands were used as input quantities together with the geometric dimensions of the rooms and measured binaural room impulse responses as described in Chapter 4. For the comparison with the measured level differences according to EN ISO 140-4, it was necessary to do auralisations for at least 5 positions in both rooms. Therefore, the binaural transfer functions in the receiving room were also measured at the same 5 positions as the level differences. For each position in the receiving room, the according standardised level differences, binaural room impulse responses, and reverberation times were used. The recordings of natural signals for the listening test were carried out for fixed

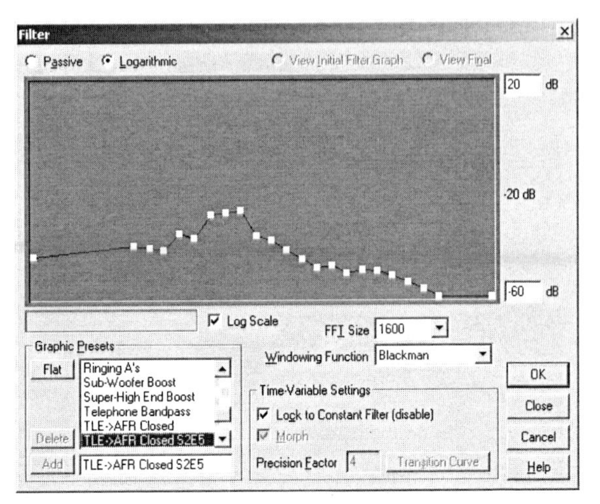

Figure 6.7: Filter dialog from audio editing software used for realisation of simple auralisation. The standardised level differences were simply fed into a filter.

positions of source and receiver. In the source room, the sound signal was recorded by a dummy head and a microphone about 10 cm above the dummy head. In the receiving room, the recording was done simultaneously with a second dummy head. The auralisation was carried out with the appropriate input data as already described in section 6.2. To verify the quality of the binaural auralisation, in parallel, a second more simple auralisation was carried out using only the standardised level differences. For it, an audio editing software (CoolEdit) was used which provides a filter function where the level differences can be adjusted over frequency (see Figure 6.7). The filter function was applied to the binaural signal measured in the source room. The resulting auralisation, then, considers only the differences in sound pressure level without simulating the reverberant sound field in the receiving room. In the following, this will be referenced to as EQ-filtering.

6.2.3 Listening Test and Psychoacoustic Evaluation

With the recordings in the receiving room and the auralised signals, a listening test was carried out as follows: The signals were presented to the subjects by headphones and replayed by a PC in paired comparisons. In a MATLAB environment, a software for a listening test was implemented. For a specific source, all paired comparisons between recording, auralisation as described in Chapter 4, and the simple auralisation as described in the previous section were to be judged. The duration of the signals was about 5 s. A list of the used sounds and their sound pressure levels can be seen in Table 6.1

Sound	Level [dB(A)]	Loudness [sone]
Clean Guitar	81	51
Distorted Guitar	85	63
Flute	74	25
Percussion	80	50
Trumpet	87	57
Vocal, Singing	70	19

Table 6.1: Sounds used as source for the listening test auralisation

In a first step, the subjects had to adjust the level of the secondly presented signal in a way, that both sounds produce the same loudness sensation. The level adjusted by the user is written into the result file and kept constant during the next two judgements. Secondly, the subjects had to choose the sound, which seems to be more 'natural', i.e., seems to be most appropriate for the given situation. From these paired comparisons, a ranking is deduced. Care has to be taken for consistent rankings, i.e., if $A > B$ and $B > C$ then $A > C$. Otherwise, a ranking can not be given. From the numbers of misjudgements, a consistency coefficient is calculated for each subject and the results of the subject are dismissed if the coefficient is below 0.6.

In the last step, the changes in the colouration had to be judged. For it, a scale ranging from 'no difference' over 'just noticeable difference', 'small difference', 'big difference' to 'very big difference' was used. In parallel, the A-weighted level and the loudness of the sound signals were evaluated by software.

6.2.4 Results

Figures 6.8 and 6.9 show the level and loudness deviation of the 2 auralised sound from the recorded sound. It can be seen clearly that there are deviations between auralisations and

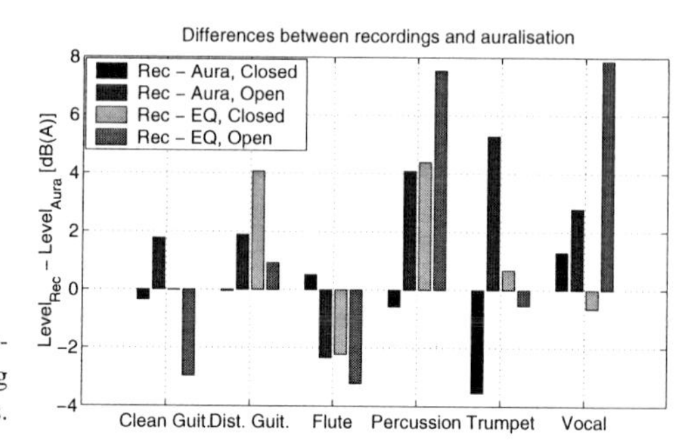

Figure 6.8: Differences in dB(A) between recordings in the receiving room and 2 different auralisations.

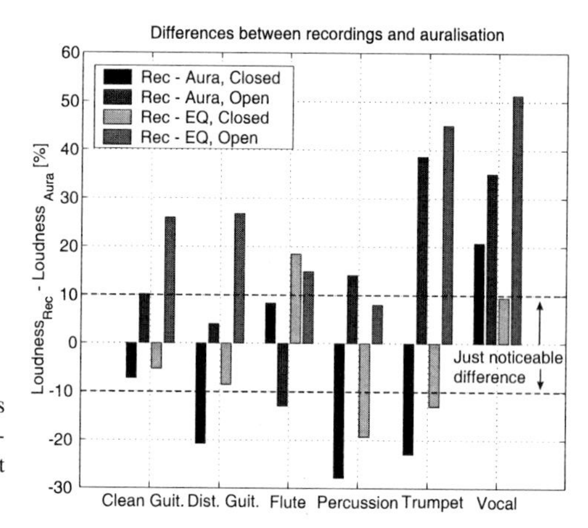

Figure 6.9: Differences in Loudness between recordings in the receiving room and 2 different auralisations.

recording. Except for one signal (Trumpet), the deviation between detailed auralisation and recording is smaller than the deviation between simple auralisation (EQ) and recording.

In Figure 6.10, the subjective level differences which were adjusted by the subjects can be seen. From the paired comparison test and the deduced ranking, a 'naturalness index' was calculated as follows: The percentage of subjects considering a signal as most natural, second natural, and least natural is multiplied by a weighting factor of 1, 0.5, and 0 respectively. The ranking of the naturalness, perceived by the subjects is displayed in Table 6.2. In the first three rows, the percentage of subjects which considered the sound as

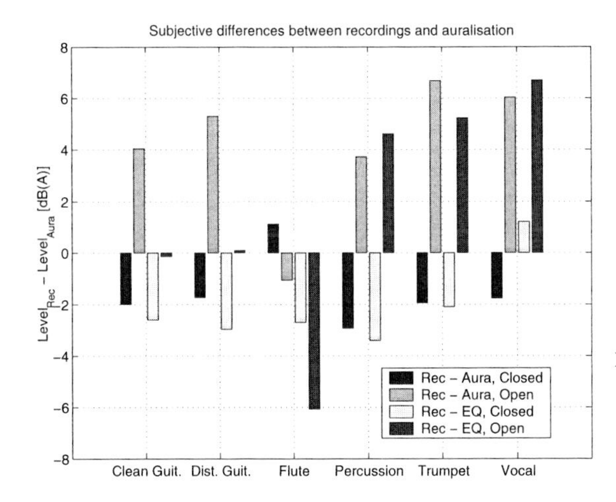

Figure 6.10: Subjective differences in level between recordings in the receiving room and 2 different auralisations

most, second, and least natural is shown. The fourth row shows the calculated naturalness index in a range from 0 to 100.

	Considered as ...	Det. Auralisation	EQ-filtering	Recording
Closed doors	Most natural	19.2	28.2	52.6
	Second natural	61.0	28.6	10.4
	Least natural	19.7	43.2	37.0
	Naturalness index	**49.8**	**42.5**	**57.8**
Open doors	Considered as ...	Det. Auralisation	EQ-filtering	Recording
	Most natural	25.3	24.1	50.6
	Second natural	48.7	28.2	23.1
	Least natural	26.0	47.7	26.3
	Naturalness index	**49.7**	**38.1**	**62.2**

Table 6.2: Calculated 'naturalness index' from paired comparison test. Maximum naturalness corresponds to a value of 100.

From the results, a clear preference for the recordings can be drawn. Further, the detailed auralisation is judged as more natural than the EQ-filtered version.

Last, the differences in colouration were judged. The difference between recording and detailed auralisation lies between 'just noticeable difference' and 'small difference'

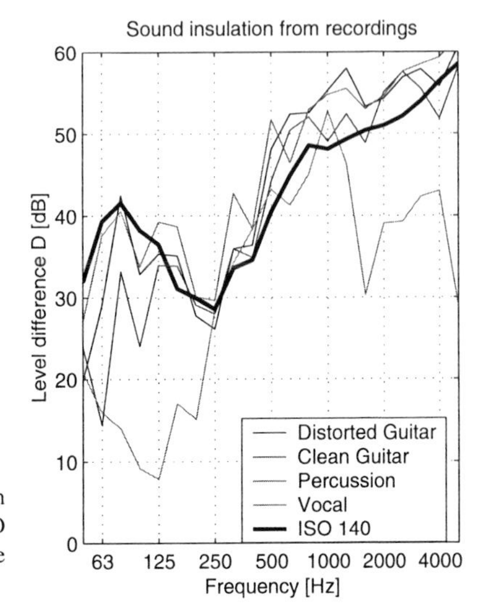

Figure 6.11: Level differences calculated from recordings and measured according to ISO 140-4. Latter ones were used for the auralisation.

whereas between recording and EQ-filtering the colouration difference lies between 'small difference' and 'big difference'. Here, also an advantage for the detailed auralisation can be seen.

6.2.5 Discussion

From the results of the listening test and the objective level measurements, it can be seen that the detailed auralisation taking the reverberation in the receiving room into account has an advantage over the simple auralisation using only the level differences. The results, however, were expected to give a more clear statement in favour of the detailed auralisation. Several reasons for deviations between recordings and detailed auralisation should be taken into account:

- The room impulse responses for the receiving room are measured with a fixed loudspeaker position. For the recordings, the 'sources' in the receiving room are given by the radiating walls. Since the reverberant part of the impulse response plays the most important role for the naturalness and colouration, this, of course, introduces deviations between auralisation and recording.

- The position of the speaker during the measurement of room impulse responses also plays an important role, especially at low frequencies. It is, thus, not possible to find a speaker position resulting in the same room impulse response as would be received when the total walls are considered as sources.

- The recordings were made with different sources (trumpet, percussion, etc.) which not always were positioned at exactly the same place as the loudspeaker used for the measurement of the level differences. Thus, the level differences for the single sources differ from the ones used for the auralisation. Figure 6.11 shows the level differences which were obtained from the recordings by subtracting the levels in source and receiving room and the one measured according to ISO 140-4. Latter ones are used for auralisation, thus it can be seen that here, already, deviations between recording and auralisation are introduced. It has to be remarked that the high deviations at low frequencies are due to the fact that the recorded signals contain very low energy in this frequency range (clearly to see for the vocal), thus, the signal to noise ratio is very low resulting in non-interpretable values.

- Despite the subjects were told to consider only the difference in colouration between two sounds, they sometimes judged differences in reverberation as differences in colouration. This falsifies the ranking between EQ-filtering and detailed auralisation, since in the EQ-filtering, no additional reverberation is introduced.

- The receiving room had, for an office room, a relatively high reverberation time (0.7 s). This might be considered as unusually reverberant by some subjects since it did not meet their expectation of the sound field inside an office room.

6.2.6 Summary

A verification of the algorithm for insulation of airborne sound insulation was carried out. Firstly, the correct reproduction of the input quantities (level differences) was proved. Secondly, the correct reproduction of the sound insulation of a real building situation was shown as satisfactory. Last, a listening test for verification of naturalness and colouration was carried out. It was shown that the perceived differences in colouration between recordings and auralisation were considered as small. The error in level reproduction compared to the real situation must be considered as slightly too high. It has, however, been taken into account that measurements of level differences in a real situation and measurements of room impulse responses can introduce errors regarding positioning of the sources and microphones. A comparison to a more simple approach which only considers the level differences between source and receiving room by one-third octave band filters shows an advantage for a detailed auralisation considering the reverberation and spaciousness in the receiving room. To get an impression of the level in the receiving room, it might be good enough to use one-third octave band filters but it must be stated that the detailed auralisation produces realistic sound impressions with almost exact spaciousness. There are clearly applications in psychoacoustic research (see Chapter 7) where a simple equalizer auralisation could not be used at all due to its rough approximation.

Chapter 7

Applications

This chapter discusses some applications which were done with the algorithm for auralisation of airborne sound insulation. The basic idea of auralisation in building acoustics was to provide a means for acoustic consultants, building engineers, or architects to get an impression of the impact of different building measures on the resulting sound impression in the receiving room. Companies producing building products should be able to give customers a sound impression besides the single number for sound insulation. Important applications besides engineering acoustics can be seen in basic psychoacoustic research. The main advantage of the algorithm is the ability to create sound material for listening tests very quickly and easily. Compared to this, to gather sound material without auralisation means to load measurement equipment (speaker, PC, microphones, amplifiers, etc.) into a car, drive to a building situation, measure the sound insulation and record signals. It can be clearly seen, that e.g. the comparison of 12 different building situation (as in the following chapter) is very time consuming if all quantities have to be measured compared to creation of sounds by auralisation.

Listening tests were carried out to demonstrate the ease of use and to investigate subjects based on perception of sounds by listeners which, normally, are hard to investigate due to the effort to create sound material. In the first section, investigations on speech intelligibility are made. In building acoustics, normally, it is desired to keep the intelligibility of speech from the adjacent room rather low to ensure a proper speech privacy. It would be desirable to calculate the speech intelligibility from the sound reduction index. Approaches to a calculating method are described.

The second chapter deals with the influence of background sounds on cognitive processes. It is based on investigations on the irrelevant speech or, more general, the irrelevant sound effect (ISE).

7.1 Investigations on Speech Perceived Through Sound Insulating Constructions

In literature, different investigations on speech intelligibility in rooms or telecommunication systems can be found. Also, different methods for calculation and quantities for characterisation of speech intelligibility were developed (STI, SII, Alcons, AI, etc.). All of these quantities have in common that the speech signal or the noise are separated into different frequency bands and the signal to noise ratios in the different frequency bands are evaluated. It is, normally, of interest to keep the intelligibility as high as possible (of course, only at justifiable costs), e.g. in theatres or conference rooms. In building acoustics, it is desirable to keep the intelligibility of speech transmitted from a source to a receiving room as low as possible to ensure speech privacy, confidentiality, and acoustic comfort, e.g. in dwellings or office rooms. Since sound insulation, usually, is described by single numbers, it is rather difficult to extract information about speech intelligibility from these quantities due to the loss of frequency dependent information. It is, therefore, desirable to calculate a single number for intelligibility from the frequency dependent sound insulation quantity and give an additional single number similar to the spectrum adaptation terms C or C_{tr}. With listening tests and parallel calculation of STI (Speech Transmission Index) and SII (Speech Intelligibility Index) values, a relation between sound insulation and intelligibility was to be evaluated. In the following, the basics about speech intelligibility, the situations and methods used for auralisation, and an approach to the calculation of quantities for intelligibility is presented.

STI	0-0.3	0.3-0.45	0.45-0.6	0.6-0.75	0.75-1
Intelligibility	bad	poor	fair	good	excellent

Table 7.1: Classification of speech intelligibility by STI values

7.1.1 Quantities for Rating Speech Intelligibility

Rating of speech intelligibility, normally, is carried out in 3 steps: separation of a speech signal or a room impulse response in different frequency bands, evaluation of a signal to noise ratio for every band, and a weighted addition of quantities describing the signal to noise ratio to obtain a single number. The resulting quantity is, then, categorised to give a clear understandable statement. As an example, in Table 7.1, the classification by STI values is given. In the experiments carried out in this work, the STI and the SII are used. Thus, a short description of these methods shall be given. For more detailed information, it is referred to [HS73], [HS85], and [ANS97].

STI The speech transmission index is based on publications by Houtgast and Steeneken ([HS73], [HS85]) who introduced the modulation transfer function (MTF). The MTF describes the loss of modulation due to the transmission over a system may it be a room or a telecommunication system or the like.

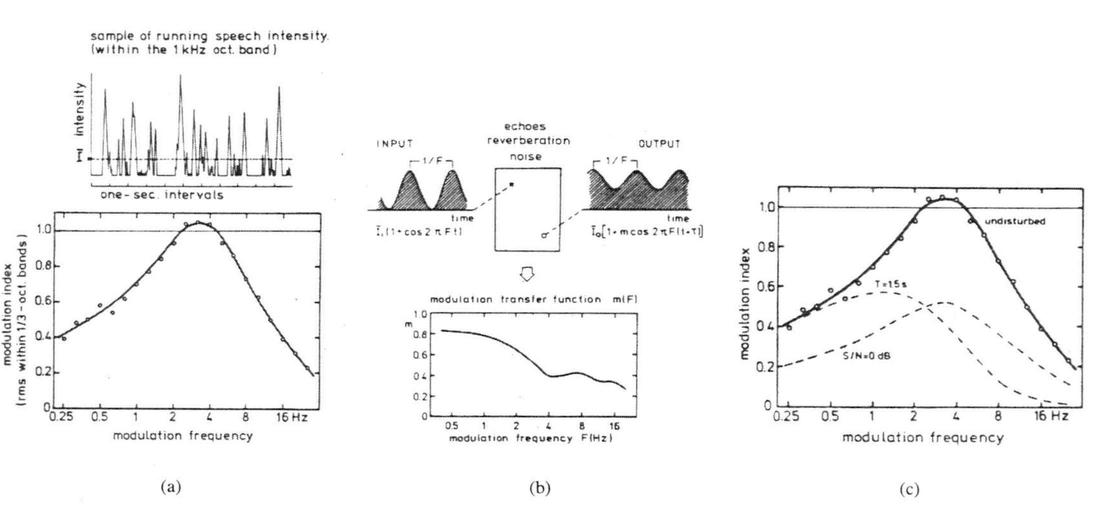

Figure 7.1: Fluctuations of the intensity envelope of speech within the 1 kHz octave band normalised with respect to the mean value (a), characterisation of a speech transmission path by the modulation transfer function $m(F)$ (b), and the effect of reverberation ($T = 1,5$ s) and noise ($S/N = 0$ dB) on the speech envelope spectrum (c). All figures taken from [HS85].

Since speech can be considered as an intensity modulated signal with modulation frequencies F between 0.25 Hz and 32 Hz, the basic idea is to replay an intensity modulated signal, record the signal at the receiver and evaluate the loss of intensity modulation for the different modulation frequencies and for different frequency bands (see Figure 7.1). Alternatively, the MTF can be evaluated from an impulse response of the system to be characterised (e.g. a room impulse response). The evaluated modulation frequencies for the STI range from 0.63 Hz to 12 Hz and the frequency bands range from 125 Hz to 8 kHz. Thus, 14 values for the loss of intensity modulation for each of 7 octaves are obtained. The STI considers both, continuous disturbing noise (e.g. due to ventilation systems) and decrease in intelligibility due to reverberation. The 14 values for the modulation transfer function (m-values) of one octave band are calculated as follows:

$$m(F) = \left\{1 + \left[2\pi F\left(\frac{T}{13.8}\right)\right]^2\right\}^{1/2} \cdot \left(1 + 10^{\frac{-S/N}{10}}\right)^{-1} \tag{7.1}$$

where F denotes the modulation frequency and T the reverberation time. The first term considers reverberation and the second term the signal to noise ratio. From the 7x14 m-values, a corresponding S/N value is calculated by

$$(S/N) = 10\log\left[\frac{m}{1-m}\right] \tag{7.2}$$

Afterwards, the resulting S/N values are limited to ±15 dB and for each frequency band, an unweighted mean value $(S/N)'_k$ is calculated from the 14 values for the modulation frequencies. Then, a weighted mean value is calculated by

$$\overline{(S/N)'} = \sum_{k=1}^{7} w_k \cdot (S/N)'_k$$

with

$$w_k = \begin{pmatrix}0.13 & 0.14 & 0.11 & 0.12 & 0.19 & 0.17 & 0.14\end{pmatrix} \tag{7.3}$$

The final value for the STI results from

$$STI = \frac{\overline{(S/N)'} + 15}{30} \tag{7.4}$$

and ranges from 0 to 1.

SII The speech intelligibility index is defined by an ANSI standard from 1997 [ANS97] and is an improvement of the articulation index (AI). For the calculation, spectra of the speech and noise signal are used in different configurations which differ in the use of frequency bands. It is, e.g., possible to use one-third octaves, octaves, or 2 different arrangements of critical bands where the bandwidth and mid-frequencies are set according to the perception of sound in human hearing. Due to the different mid-frequencies, bandwidth, and the importance of a specific band for the intelligibility, different band weighting factors I_i result. The sum of the I_i values is always 1.

In this work, one-third octaves are used since most of the used data (sound insulation, speech spectrum, etc.) are given or to be measured in one-third octaves. 18 one-third octave bands between 160 Hz and 8000 Hz are, then, considered. The band weighting factors I_i are shown in Figure 7.2 (right). From this, it can be seen that the most important frequency bands for good speech intelligibility lie around 2 kHz.

The SII is calculated by the band weighting factors and the band audibility function as follows

$$\mathrm{SII} = \sum_{i=1}^{18} I_i \cdot A_i \tag{7.5}$$

The procedure for calculating A_i shall not be explained in detail. It can, however, be summarised that standardised spectra for speech at different levels, the hearing threshold, spectra of the background noise, and the self-speech masking level are considered. Figure 7.2 (left) shows the speech levels, standardised in [ANS97] for different speech styles. These spectra result from averaging the spectra of several male and female speakers.

For the evaluations in this work, own speech material spoken by only one male speaker was used. The spectrum of this speech material was measured and used instead of the standardised spectra. Details on the measurement procedures can be found in [Hol03]. The results of the measurements are shown in one of the following passages.

7.1.2 Listening Tests and Measurements

First experiment In a first experiment, the correlation between the single number for rating sound insulation and speech intelligibility was investigated. 6 different room situations

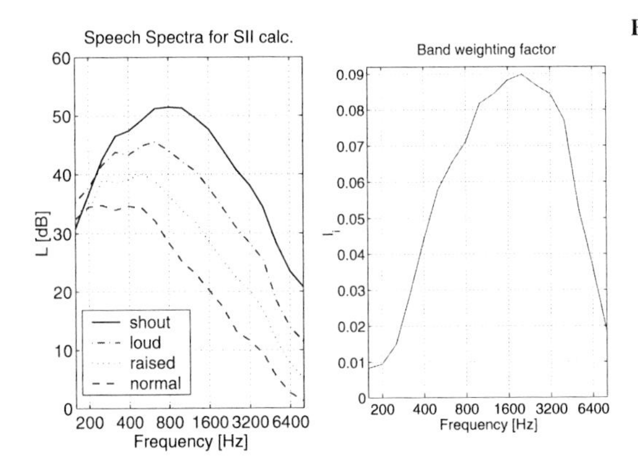

Figure 7.2: The left figure shows speech spectra for the SII calculation for different speech styles, standardised in [ANS97]. Note that spectra and not levels are displayed, i.e., the levels are divided by the bandwidth. In the right figure, band-weighting factors for the 18 one-third octave bands are shown which represent the importance of the frequency bands for speech intelligibility.

	CAC 1	CAC 2	CAC 3
Required R'_w	53	56	59
Loud speech is	intelligible	generally intell.	generally not intell.

Table 7.2: Classes of acoustical comfort and their definitions according to VDI 4100

were created, the sound signals in the receiving room were auralised and used for listening tests (see Figure 7.3). The constructions were chosen in a way that each two of them had the same weighted apparent sound reduction index R'_w and belonged to a class of acoustic comfort according to VDI 4100. In VDI 4100, 3 different classes of acoustic comfort are described by requirements for sound insulation and properties like speech intelligibility, etc. The requirements for class 1 are the same as defined in the German standard DIN 4109 [DIN89] (R'_w = 53 dB between dwellings). For class 2 and 3, the requirements for R'_w are raised by 3 dB each. Also, the quality of speech intelligibility for certain speech levels is defined (see Table 7.1).

For the listening tests, material from a sentence test for speech intelligibility was used as source signal. It consists of simple sentences with 5-6 words each. The auralised signal in the receiving room was presented to the listeners by means of a CD-Player and electrostatic earphones (STAX) in an anechoic room to prevent disturbance from environmental noise. To simulate environmental noise in a dwelling and to have a defined noise level, pink noise

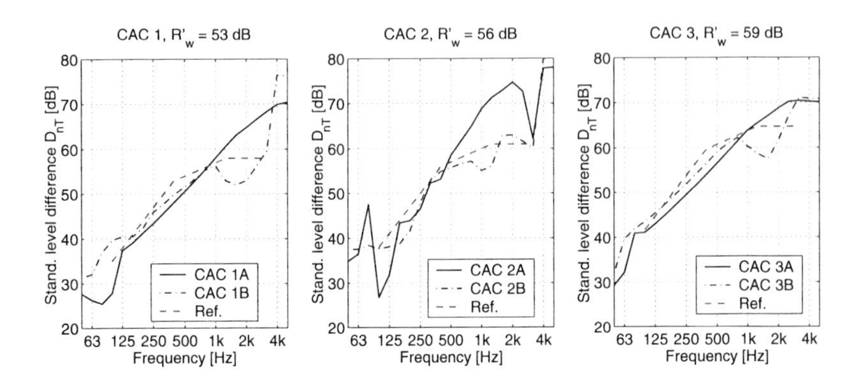

Figure 7.3: Standardised level difference for 6 room situations. Each two of them have equal weighted apparent sound reduction indexes R'_w according to the classes of acoustical comfort in VDI 4100.

at 20 dB(A) was added to the auralised signals. The speech level in the source room was adjusted to be about 80 dB(A) (loud speech). The 17 test persons were mostly recruited from the staff of the institute. They had to listen to the speech material and repeat the words they understood in a pause after each sentence. The number of sentences per room situation was 30 with about 150 words. The speech intelligibility was derived by dividing the number of understood words by the total number of words. A first test sequence consisted of 20 sentences, thus 200 sentences were presented in total. The whole test lasted about 20-30 minutes and was considered as slightly too long by only one subject.

Results Figure 7.4 shows the results for the speech intelligibility for the 6 constructions. It can be seen that the speech intelligibility varies between situations with equal R'_w (inside the same class of acoustic comfort) and also that for higher classes of acoustic comfort, the speech intelligibility does not necessarily decrease. From this, it can be concluded that there is no linear relation between the single number rating for sound insulation and speech intelligibility or speech privacy. Thus, a quantity must be derived from the frequency dependent quantity. The speech transmission index (STI) was calculated from the spectra of

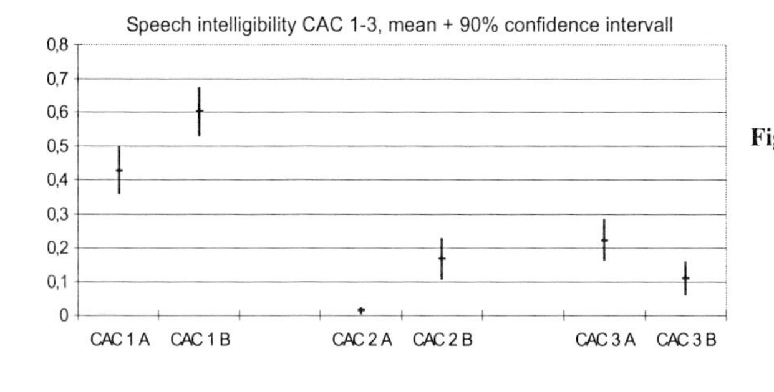

Figure 7.4: Resulting speech intelligibility from listening tests with 6 room constructions according to classes of acoustical comfort according to VDI 4100.

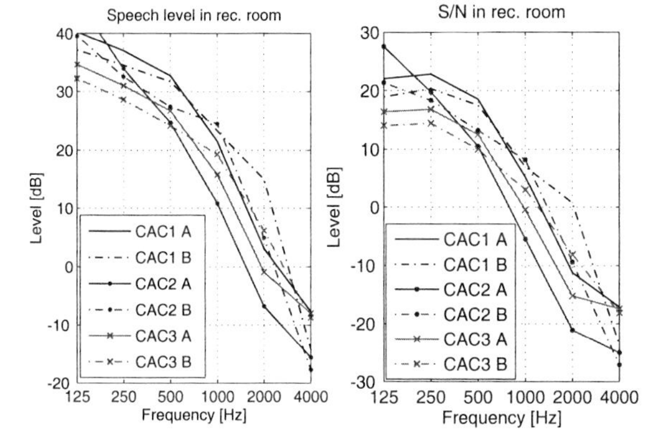

Figure 7.5: Speech level (left) and signal to noise ratios (right) in the receiving room.

speech and noise. To consider the reverberation, in a second step, a reverberation time of 0.5 s was assumed which was approximately the reverberation time of the receiving room at 500 Hz. The speech and noise spectra as well as the resulting signal to noise ratios in the receiving room can be seen in Figure 7.5. The results for the speech transmission index compared to the intelligibility resulting from listening tests are shown in Figure 7.6. It can be seen that the relation between results from listening tests and calculation do not match very good. The decision which situation results in a better speech intelligibility is the same for STI values and intelligibilities from listening tests only for situations CAC1 and CAC2. For CAC3, the STI gives a different result. To get more information on this, more listening tests with different situations were to be carried out and also some improvements on the

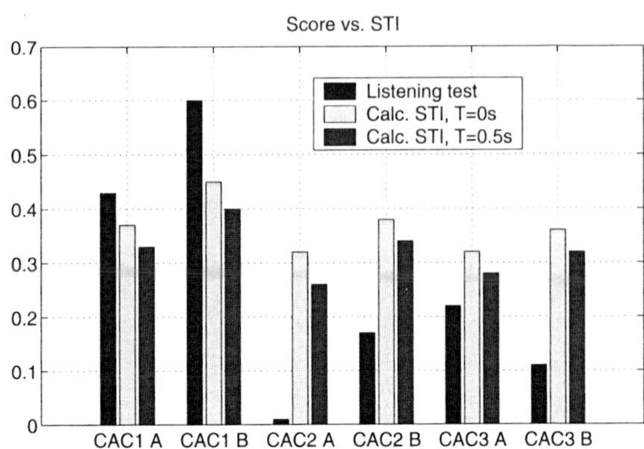

Figure 7.6: Speech intelligibility from listening tests (blue bars, left) compared to the speech transmission index calculated from speech and noise spectra (green, middle) and, additionally considering a reverberation time of 0.5 s(red, right).

listening test set-up were to be done.

Further Studies In a second step, 12 different room situations were auralised and listening tests were carried out. In a preliminary test, the speech intelligibility of sounds auralised by the algorithm described in Chapter 4 was compared to a simple auralisation by just filtering the source sounds with the level differences as described in section 6.2.2, pp. 79. The results are shown in Table 7.3. It can be seen that the complete auralisation with reverber-

Intelligibility		
	EQFilter	Aura
Group 1	0.98	0.80
Group 2	0.94	0.77
Total	0.96	0.79

Table 7.3: Results of preliminary test comparing simple and complete auralisation

ation in the receiving room causes a decrease in speech intelligibility. This, of course, is due to the fact that more reverberation is introduced compared to a simple application of a filter and corresponds more to real conditions. With the experiences in the preliminary test, several improvements over the first experiments were introduced:

- The speech level in the source room was adjusted to 65 dB (A) which seems to be

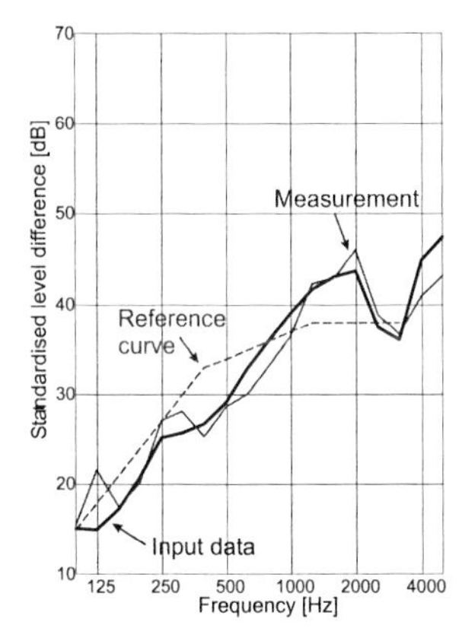

Figure 7.7: Comparison of the input data for auralisation and the resulting level difference reproduced via crosstalk-cancelled loudspeakers after auralisation.

more realistic compared to 80 dB (A) in the first experiments.

- Audio signals were replayed via loudspeakers with crosstalk cancellation (see Chapter 3.3, p. 38). This makes measurements of the sound pressure level at the ears of the listener more accurate. For the SII, the level has to be measured between the ears when the listener is absent. When using headphones, the level has to be measured with an artificial ear and corrections to obtain the free field level have to be applied which can be found in the standard describing the SII. Unfortunately, these corrections are not valid for the STAX headphones which were used in the tests. Using cross-talk cancellation, the level can be measured by an artificial head and corrected by the free-field transfer function.

- It was made sure by another preliminary tests that the room situations provide a speech intelligibility *between* 0 and 1. If all situations provide a very low or very high intelligibility, no information can be won.

Verification In a first verification, the level differences reproduced by the auralisation and presentation via crosstalk-cancelled loudspeakers were compared to the input data. From the result in Figure 7.7, it can be seen that the input data is quite well reproduced.

A second verification of the auralisation was done by using a real situation (2 rooms in the Institute of Technical Acoustics with rather low sound insulation). The level differences, binaural room impulse responses and reverberation times were measured and calculated the same way as illustrated in section 6.2.1. The listening test consisted of two presentations: in the first, the subjects were placed in the real receiving room and listened to speech which was replayed by an audio system in the source room. They had to repeat the understood words and the intelligibility was derived from the number of understood words divided by the number of total presented words just as described above. The second listening test was carried out with signals resulting from an auralisation of the room situation. The auralisation was carried out using the appropriate input data as described in section 6.2.2. The auralised signals were replayed via loudspeakers in an anechoic room using crosstalk-cancellation. The speech intelligibility of the two tests should ideally be the same if the auralisation works correctly. For the listening test in the real situation, a speech intelligibility of 0.56 was obtained and for the auralised signals the value was 0.53, i.e. 53% and 56% of the words were understood which is a good result.

Listening Tests Then, listening tests for speech intelligibility of 12 room situations were carried out in 2 runs with 6 situations each and different subjects. This was due to the number of test sentences which were only 200 and it was considered to have at least 30 sentences per situation. Figures 7.8 and 7.9 show the standardised level differences D_{nT} for the 12 situations.

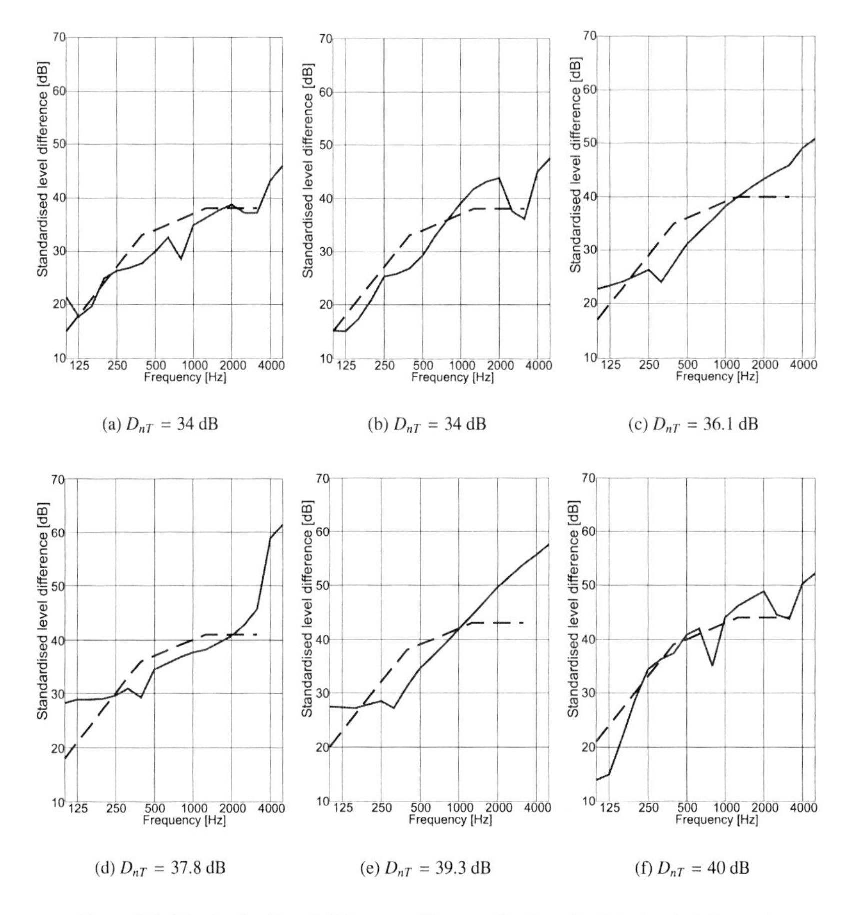

(a) $D_{nT} = 34$ dB

(b) $D_{nT} = 34$ dB

(c) $D_{nT} = 36.1$ dB

(d) $D_{nT} = 37.8$ dB

(e) $D_{nT} = 39.3$ dB

(f) $D_{nT} = 40$ dB

Figure 7.8: Standardised level difference of 6 room situations for listening tests in run 1.

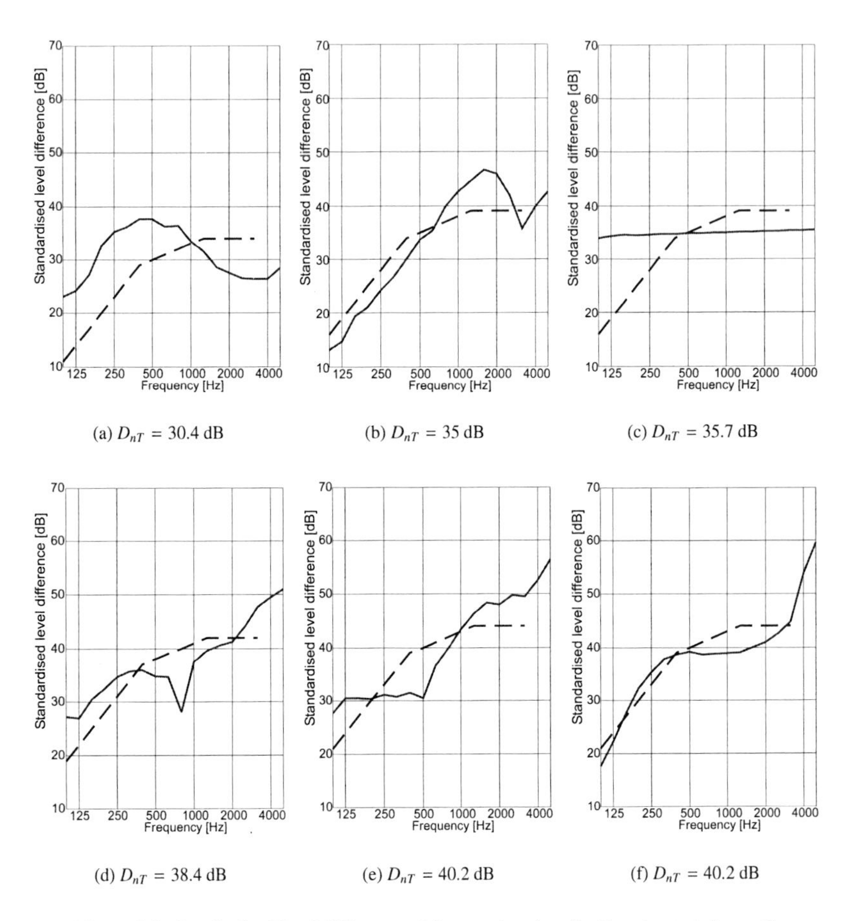

(a) $D_{nT} = 30.4$ dB (b) $D_{nT} = 35$ dB (c) $D_{nT} = 35.7$ dB

(d) $D_{nT} = 38.4$ dB (e) $D_{nT} = 40.2$ dB (f) $D_{nT} = 40.2$ dB

Figure 7.9: Standardised level difference of 6 room situations for listening tests in run 2.

Results and Discussion Table 7.4 and Figure 7.10 show the results for the intelligibility derived from the listening test and the calculated values for the STI. Also 2 values for the SII are shown which are calculated using band-weighting factors for a speech test (test) and for the one-third octave based method (1/3 oct.). It can be seen that there are deviations

$D_{nT,w}$ [dB]	Score	STI	SII (test)	SII (1/3 oct.)
30.4	0.72	0.58	0.74	0.71
34.0(1)	0.81	0.46	0.65	0.56
34.0(2)	0.80	0.42	0.58	0.48
35.0	0.51	0.38	0.53	0.43
35.7	0.53	0.47	0.64	0.58
36.1	0.82	0.42	0.54	0.44
37.8	0.73	0.43	0.54	0.45
38.4	0.49	0.41	0.53	0.44
39.3	0.54	0.36	0.43	0.33
40.0	0.50	0.34	0.42	0.33
40.2(1)	0.46	0.35	0.44	0.33
40.2(2)	0.35	0.39	0.48	0.41

Table 7.4: Results of listening tests for speech intelligibility of 12 situations

between the results for the intelligibility and the calculated values. It must, however, be noted that the calculated STI values have to be re-transformed into values for intelligibility to be compared to the other numbers. For the transformation, a relation between STI and intelligibility has to be found. This, however, is dependent on the speech material and listening test conditions. Figure 7.13 shows some transformation curves for different kinds of speech material. It can be seen that for less semantic content, the curve goes more and more to a straight line between intelligibility and STI. The more semantic content, e.g., for a test with simple sentences, the more are the subjects able to guess the words which were not understood. If they, e.g., understood 'Good morning, ... gentlemen.', it is easy to guess the missing two words. Bradley [Bra86] compared lots of investigations on the STI and speech intelligibility. In Figure 7.11 the results of various listening tests versus the calculated STI is shown. From the data, a regression curve is calculated. It can be seen that there are strong deviations from the regression curve which represents a mean value for the used speech material and listening test conditions. To find a regression curve for

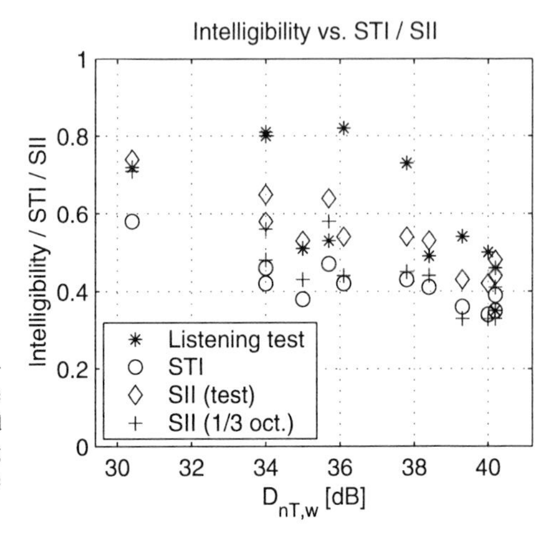

Figure 7.10: Results of the intelligibility from the listening tests, the calculated STI, and the calculated SII with different bandweighting factors for a speech test and the 1/3 octav band method.

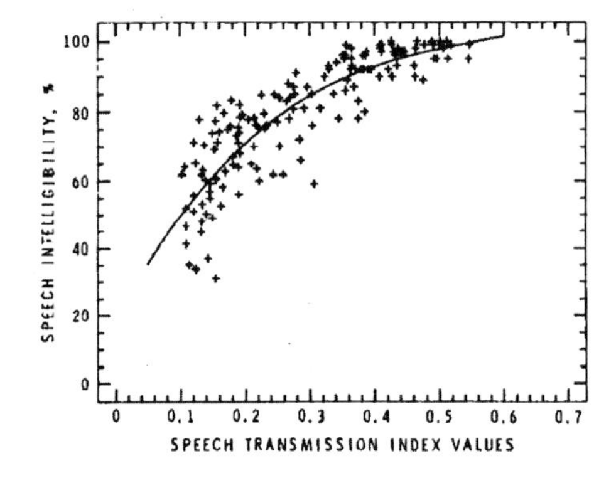

Figure 7.11: A collection of relations between speech intelligibility and STI from various investigations and a regression curve representing the mean value for different speech material and listening test conditions. (From [Bra86])

Figure 7.12: Relation of modulation perception over sound pressure level. At lower levels, the degree of modulation must be higher to be perceived by a listener. (From [ZF99]

the relation between the STI and the single number rating for the sound insulation, many listening tests have to be carried out which was not in the scope of this work. Other reasons for the deviations may be:

- The STI was developed for undistorted speech at 'normal levels' where the intelligibility in quiet is already good. In this case the speech level is just above the hearing threshold. As speech is considered as an intensity modulated signal and the most important parts of this modulation for intelligibility range from -18dB and +12 dB around the rms speech level, a great portion of the signal falls below the hearing threshold.

- The perception of modulation decreases from levels around 30 dB to the hearing threshold as can be seen in Figure 7.12. Thus, for speech signals at a very low levels, a non linear behaviour can be assumed with a faster decrease of intelligibility.

- The speech signal in the receiving room is strongly distorted by the frequency shape of the sound insulation. As one can see, the S/N ratios range from around +30 dB at 125 Hz to -30 dB at 4 kHz. It must be investigated if the STI calculation is valid under this circumstances.

- The relation between intelligibility measured in sentence tests and the STI shows a steeply falling slope at lower STI values, so only a little change in STI can come along with great variations in the corresponding intelligibility.

- The subjects in the test concentrated very much on understanding the words, so the results from the listening tests seem a bit too high.

The same is true for the relation between SII and sound insulation. Here, bandweighting factors suitable for the used listening test have to be evaluated. This, however, would also make it necessary to gather data by extensive listening tests with lots of different building situations. If, in the future, it is desired to establish a single number rating for speech intelligibility, this could be one way of realising it. Another way would be, to do auralisations with different sound insulation spectra, use the speech spectrum provided by the SII

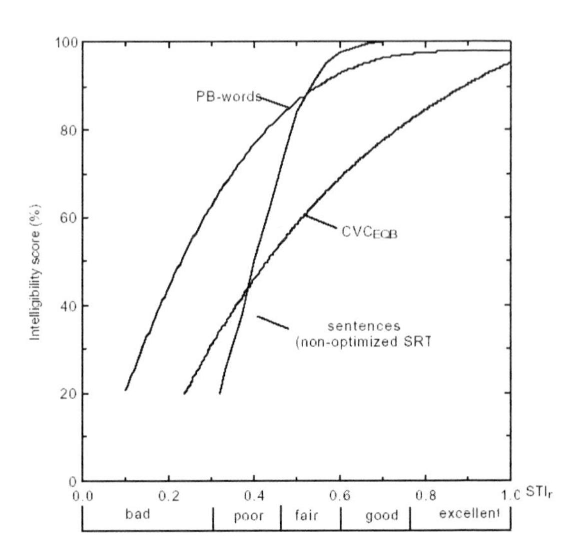

Figure 7.13: Relation between the score (result of listening tests for intelligibility) and the STI value of different speech material and types (PB-Words: phonetically balanced words, CVC: consonant-vowel-consonant). Below the x-axis the classification scheme for the STI is shown. It can be seen that for less semantic content (sentences over PB-words to CVC) the curve goes towards a straight line. From [HS84]

standard and calculate the SII from the gathered data to fill Figure 7.10 with more results to obtain a regression curve which represents a relation between single number rating and SII.

Building acoustical auralisation can be considered as a helpful tool, not only to find relations between single number ratings of sound insulation and speech intelligibility but for any relation between a desired parameter and sound insulation.

7.2 Investigations on the Disturbance of Speech Perceived Through Sound Insulating Constructions

At the Institute of Work, Environmental and Health Psychology at the Catholic University of Eichstätt-Ingolstadt together with the Institute of Technical Acoustics at RWTH Aachen University, the Irrelevant Speech Effect (ISE) was investigated. The ISE describes the influence of irrelevant background speech on verbal short-term memory performance and should be considered, e.g., in open-plan offices or classrooms. The content of speech is irrelevant for the task.

An experiment for investigating the effect of different background sounds typically can be described as follows: On a screen, a randomised series of numbers from 1 to 9 are presented to a subject in 9 seconds. The digits are visible for 0.75 seconds and after a pause of 0.25 seconds where no digit is visible the next one is presented. After another 10 seconds, the subject has to enter the series of numbers in the order as displayed using a mouse. The total error rate and the errors per position in the series are evaluated. During the experiment, different background sounds are presented which have no relation to the task. In former investigations, it was found that the intelligibility of background speech has nearly no influence on the performance since the error rate of the test was almost equal for German and Japanese speech (with German subjects) and for reversed speech signals (see overview article from Klatte and Hellbrück [KH93]). Also, no influence of the level of speech between 40 and 76 dB was found. It was, however, found that sounds different from speech could also cause errors. It can be said that sounds which are similar to speech in amplitude and frequency modulation cause more errors. For example, legato music (chants of Gregorian monks) does not cause significantly more errors than the control condition (pink noise or silence) but staccato music (flute) causes significantly more errors.

In a first experiment, four different sounds were presented as background: Speech in the source room at 55 dB(A), auralised speech in the receiving room at 35 dB(A) but with different speech intelligibilities due to different shapes of the sound insulation curves, and pink noise at 25 dB(A). For the experiment, 2 different room situations were created and auralised. The corresponding curves of the standardised level difference D_{nT} can be seen in Figure 7.14. The sound insulation was chosen in a way, that the loudness of the auralised sounds was at the same value, but, the speech intelligibility was rather different. This was proved by a listening test where subjects had to recognise sentences of auralised speech presented by headphones. These tests are described in detail in section 7.1. The intelligibility is expressed by the ratio of understood words to the total number of words. For the 2 situations, the intelligibilities were 98% (Speech_35G) and 75% (Speech_35B). It has

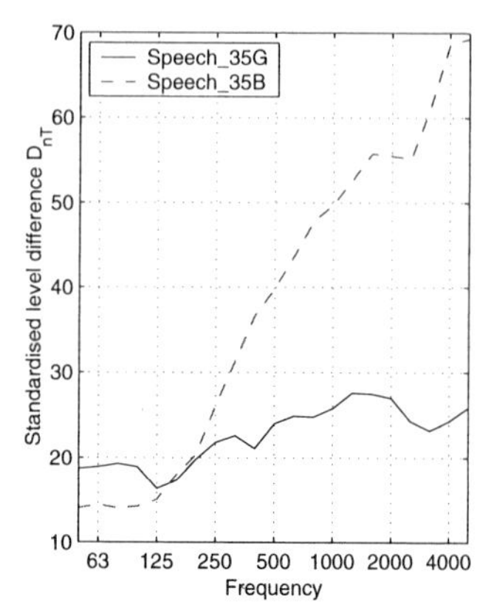

Figure 7.14: Standardised level difference D_{nT} of the 2 room situations used for the ISE experiment. Speech_35B (dashed) causes a relatively poor intelligibility due to the steep slope.

to be remarked that the test material consists of german sentences with 5-6 words which are easy to understand. The subjects were in some cases able to reconstruct missing words from the rest of the sentence. Thus, an intellibility of 75% is rather poor. First results show that there is a significant difference between the performances for the two auralised signals at 35 dB(A) with different intelligibilities and also between the speech in the source room and the speech with bad intelligibility, *but not* between the source room speech at 55 dB(A) and the speech at 35 dB(A) with good intelligibility (Schlittmeier, Thaden [ST]).

From this first experiment, the conclusion could not be drawn that it is the speech intelligibility that matters and not the level. In a second experiment, speech intelligibility and content of speech are disentangled by using Japanese speech. If the speech intelligibility is responsible for the different results between the auralised signals with good and bad intelligibility, then there should be no difference between these two signals for a Japanese speaker since the intelligibility is zero by default. To keep the speech spectrum closely to the one used in the first experiment, the signal with Japanese speech was filtered to have the same long time spectrum as the signal with German speech. Then, an auralisation with the same parameters as in the first experiment was carried out.

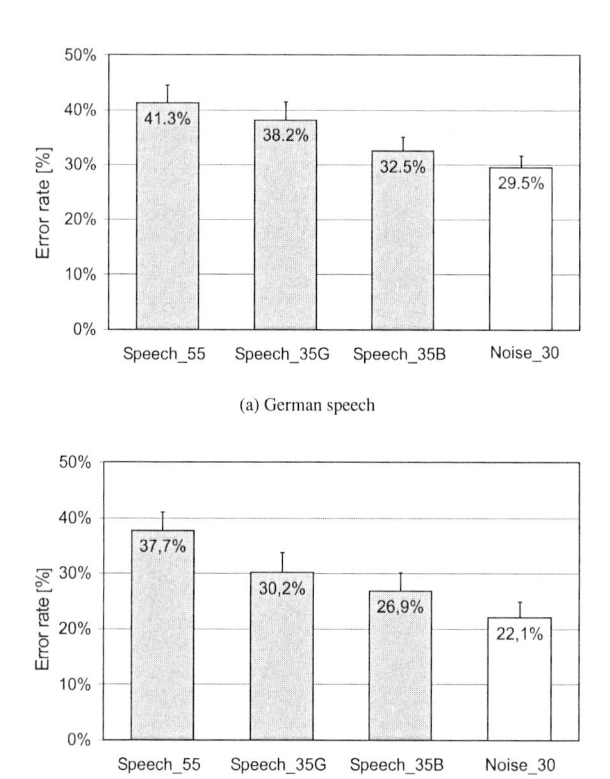

(a) German speech

(b) Japanese speech

Figure 7.15: Total error rates for the ISE test with German and Japanese speech at 55 dB(A), auralised speech at 35 dB(A) with different speech intelligibilities and pink noise at 30 dB. Mean value and standard deviation.

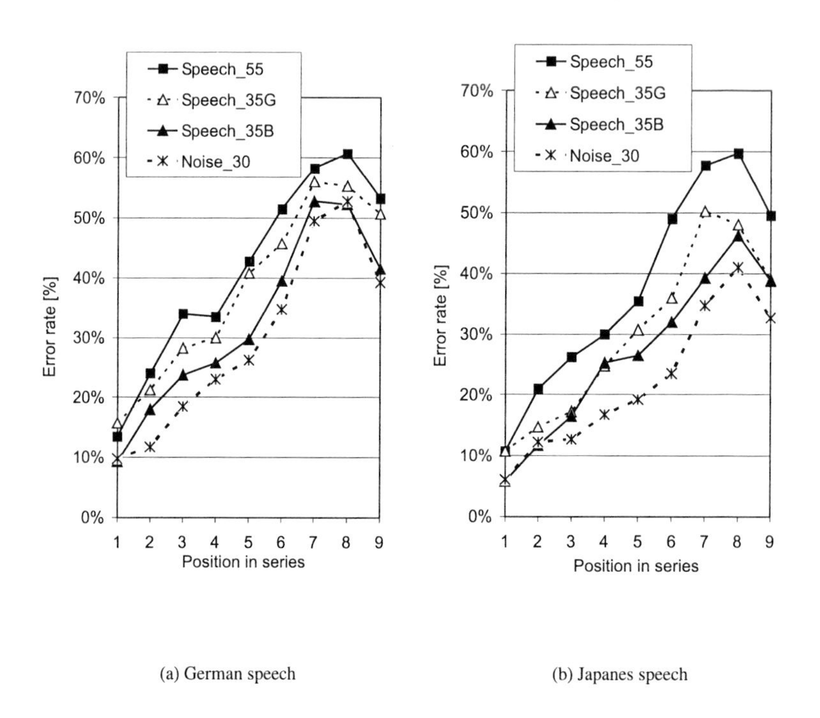

(a) German speech

(b) Japanes speech

Figure 7.16: Errors depending on the position of the number in the series for German and Japanese speech at 55 dB(A), auralised speech at 35 dB(A) with different speech intelligibilities and pink noise at 30 dB.

The results from the performance tests can be seen in Figure 7.15 (total error rate) and Figure 7.16 (errors depending on the position in the series of digits). Additionally, the subjects were asked to answer a questionnaire how difficult, demanding, annoying, disturbing they perceived the task under the different background sounds and how they judged their concentration. The scale ranged from 0 (not at all) over 1 (hardly), 2 (mediocre), and 3 (rather) up to 4 (extraordinary). Figure 7.17 shows the results for the questionnaire.

From these preliminary results and detailed statistical evaluations, a secure conclusion can not be drawn. It seems that speech intelligibility plays a role for the differences between the two auralised sounds at 35 dB(A) since these differences vary between German and Japanese speech. This is contradictory to the investigations about the irrelevant speech effect which lead to the conclusion that intelligibility does not matter regarding the disturbance of background sounds. Detailed investigations on the modulation spectrum of the sounds and more room situations/sound insulation curves have to be carried out. From these investigations it is, then, possible to derive room and building acoustical measures to improve working conditions in offices, e.g., by focussing on reducing intelligibility. One measure could be to increase background noise by ventilation systems or to introduce artificial background noise since this decreases intelligibility due to the lower signal to noise ratio and is not considered as disturbing as speech. The building acoustical auralisation proved as a helpful tool for these kinds of investigation.

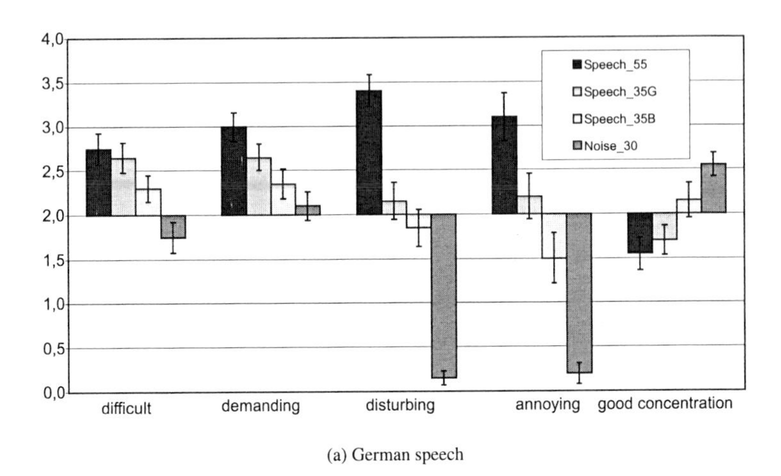

(a) German speech

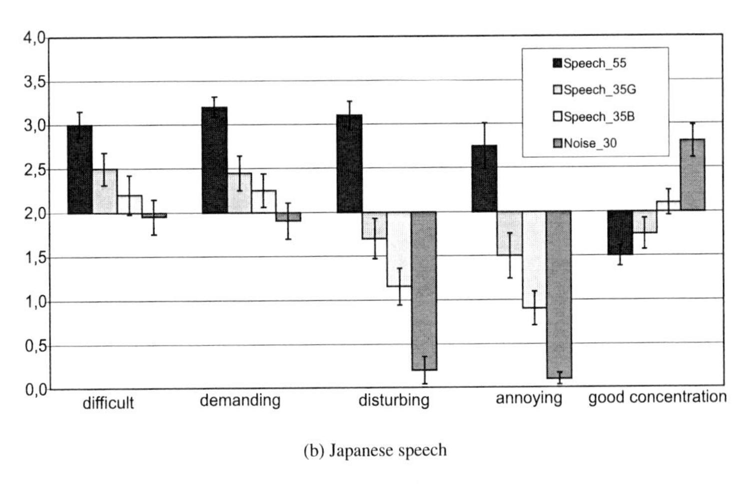

(b) Japanese speech

Figure 7.17: Subjective evaluation of different parameters for the listening test with German and Japanese speech from a questionnaire.

Chapter 8

Summary

In this thesis, first, the importance for a building acoustical auralisation is presented. Due to increasing noise problems, and the consideration as noise as a threat for human health and social life, an improvement of noise protection has to be acchieved. Since building acoustical quantities are, normally, handled by single numbers which are mostly not familiar to 'normal' people, an auralisation of the effects of different building materials, building styles and measures to improve the insulation of already constructed dwellings is helpful to give dwellers an illustrative listening experience. Also for research purposes like listening tests with lots of different building situations to find, verify, or improve, e.g., a suitable rating scheme for sound insulation, the auralisation is useful. In Chapter 2 prediction models for the calculation of sound insulation for a building situation from product data are presented. First, the application of (many different) rating quantities and requirements in Europe are shown which seems slightly confusing. Then 2 prediction models, the statistical energy analysis and the calculation model according to the European standard EN 12354, are briefly explained. From the results of these calculations, a building acoustical auralisation can be carried out. Chapter 3 illustrates the basic principles of auralisation regarding signal processing, binaural techniques, and issues of presenting signals by different means and the advantages and drawbacks of different presentation techniques.

An algorithm for the auralisation of airborne sound insulation is presented in Chapter 4. The algorithm was developed using predicted input quantities as the standardised level difference and binaural room impulse responses. A simplified model for the sound transmission and the sound field in the receiving room was used to obtain an authentic but not physically exact sound field. The single processing steps and the simplifications made in the approach are explained. The algorithm was implemented on a PC and combined with a software for calculation sound insulation according to EN 12354.

Chapter 5 deals with the auralisation of impact sound insulation. First a simplified algorithm for calculating the sound field in the receiving room from the impact sound level, a measured force source and information about the receiving room is presented and verified regarding the reproduced levels. Since the interaction between a source and the structure plays an important role, the characterisation of sources was the main task to be solved. The mechanical impedance and force of a source had to be determined by an appropriate measurement method and setup. A method for determining these quantities is presented and verified. Different approaches for suitable measurement setups are tested, all of them produced reasonable results for special kinds of sources (modified tapping machine, rubber ball), but due to mechanical problems and difficulties, a setup for the determination of the impedance of arbitrary force sources is not yet finished.

Chapter 6 concentrates on the verification of the auralisation of airborne sound insulation. First, the reproduction of the level differences used as input is verified by measuring the levels of the auralised sounds with a sound level analyser and calculating the level differences from the auralised sounds. It shows that the auralisation almost perfectly reproduces the level differences. Additionally, a real building situation was used for a comparison of measured and auralised level differences and recorded and auralised sounds. Building acoustical measurements according to ISO 140-4 and room acoustical measurements of the reverberation times and binaural room impulse responses were carried out. These quantities were used as input for an auralisation. The same positions were used for measurements and auralisation, thus, a measurement according to ISO 140 from the auralised sounds was possible. The measured level differences were produced almost perfectly by the auralisation.

A listening test was carried out to prove the 'naturalness' and level sensation compared to dummy head recordings in the receiving room. There are slight deviations in level which may be explained by uncertainties in measurements and recordings. In total, however, the airborne sound auralisation was verified correctly.

In Chapter 7, applications for the airborne sound auralisation are presented. One section concentrates on the perception of speech sounds through sound insulating constructions. An approach to the finding of a rating scheme for speech intelligibility depending on the sound insulation of a construction is presented. Listening tests with speech signals were carried out for different building situations with varying sound insulation. The reproduction of the correct speech intelligibility was verified by a comparison of auralised sounds to sounds recorded in a real building situation. Since the finding of a rating scheme would include lots of listening tests which are beyond the scope of this work, this may be picked up by research groups and committees, dealing with rating schemes in the future.

In a second application, the disturbance of background speech for mental processing is investigated. Listening tests with different auralised speech signals were carried out to investigate the relation between intelligibility, level, and frequency curves of speech signals and the disturbance of mental work. This work was done in cooperation with the Catholic University of Eichstätt.

Summarising the results of this thesis, it can be said that

- An algorithm for auralisation of airborn sound insulation has been developed, implemented, and verified

- The validity and usefulness of this algorithm has been shown in several applications

- An approach to an auralisation of impact sound was presented

- A simplified auralisation of impact sound was developed, implemented and verified regarding level reproduction

- A method for characterising impact forces was developed and verified

- Different approaches for measurement setups to characterise impact forces were implemented. They worked for special force sources but due to mechanical problems and difficulties it was not yet possible to find a measurement setup for arbitrary force sources.

- A verification of the algorithm for airborne sound insulation was carried out by a comparison of auralisation and recordings in a real building situation
- Applications in practical acoustics, consulting, etc. is now possible
- Applications of airborne sound insulation on speech perceived through sound insulating constructions were presented. This includes an approach to a rating scheme for speech intelligibility (or speech privacy) from the frequency dependent sound insulation and investigations on the disturbance of background speech perceived through sound insulating constructions (e.g. offices).

Outlook Since the algorithm for auralising airborne sound is mostly completed and verified, only slight improvements can be introduced here. With PCs gaining more computational power, the building acoustical auralisation can be combined with room acoustical simulations of the sound field. The improvements due to this, however, are considered as small. More applications of the algorithm may be carried out, e.g., verification of reference curves for single number ratings by listening tests.

The impact sound insulation lacks the availability of valid input data as the characterisation of force sources is not yet finished. A promising approach for measuring source impedances and forces may be in the use of light-weight composite materials. Future work should concentrate on realising a measurement setup. It is to investigate, how additional layers on a floor construction should be considered by a model for impact sound transmission since the result of a combination also depends on a dynamic interaction between floor and covering.

Also, in future a combination of airborne and impact sound auralisation is desirable since the sound produced by a walker in the source room also is transmitted via airborne sound.

Kapitel 9

Kurzfassung

Die Numerierung der Abschnitte in dieser Kurzfassung entspricht der Numerierung der Kapitel der Dissertation. Wegen der Kürze des Textes ist auf die Darstellung von Bildern verzichtet worden, daher sei zur Erläuterung der Abschnitte auf die Darstellungen in den entsprechenden Kapiteln der Dissertation verwiesen.

Einleitung

Die Schalldämmung einer Wohnung gegen Außenlärm sowie Innenlärm spielt eine große Rolle für den akustischen Komfort und damit für den Wohnwert. In Deutschland ist der Bauherr verpflichtet, einen Nachweis des Schallschutzes nach DIN 4109 zu führen. Dazu werden momentan nach Beiblatt 1 der DIN 4109 Tabellen und Daumenregeln angewandt, um aus Daten einzelner Bauprodukte die Schalldämmung der Gesamtsituation zu berechnen. Es werden dazu Einzahlkennwerte für die Schalldämmung der einzelnen Bauprodukte verwendet. Ein großer Nachteil der Einzahlkennwerte ist die geringe Aussagekraft. Weder ein akustischer Laie noch ein Fachmann kann sich eine Vorstellung von der Schalldämmung einer Raumsituation machen anhand eines Einzahlkennwertes wie des bewerteten Schalldämmaßes R'_w von z.B. 53 dB. Zudem ist die Abbildung der frequenzabhängigen Schalldämmung auf einen Einzahlwert nicht eindeutig, so dass unterschiedliches Frequenzverhalten zum gleichen Einzahlkennwert führen kann. Im Rahmen der Arbeit wurde ein Algorithmus zur Auralisation (Hörbarmachung) der Schalldämmung in Gebäuden entwickelt, verifiziert und in unterschiedlichen Anwendungsgebieten erprobt. Das Verfahren der Auralisation ist in der Raumakustik bereits etabliert, um sich einen Eindruck z.B. der akustischen Qualität eines Raumes (Konzertsaal, etc.) zu verschaffen. Dieses lässt sich durch die Darbietung des Signals viel leichter erreichen als durch die Betrachtung vieler Einzahlkennwerte wie Nachhallzeit, Seitenschallgrad, etc. In der Bauakustik sind bisher nur wenige Anwendungen bzw. Entwicklungen der Auralisation bekannt.

Vorhersagemethoden in der Bauakustik

Die Methode zur Vorhersage der Schalldämmung ist in den europäischen Ländern nicht einheitlich geregelt. Es existieren insgesamt 10 unterschiedliche Konzepte für die Angabe der Luftschalldämmung mit unterschiedlichen Grenzwerten sowie 6 unterschiedliche Konzepte bei der Angabe der Trittschalldämmung. Häufig basieren die Vorhersagemethoden auf Einzahlkennwerten der verwendeten Bauprodukte. Das in Deutschland verwendete Verfahren, welches auf Korrekturtermen und Tabellen basiert, wird kurz beschrieben. Anschließend werden 2 Verfahren, welche auf frequenzabhängigen Eingangs- und Ausgangsdaten basieren, beschrieben, nämlich die statistische Energieanalyse und das Verfahren nach EN 12354. Ersteres modelliert die Übertragung von Schall über Trennwände durch aneinander gekoppelte Oszillatoren mit Verlust- und Kopplungsfaktoren (Wände, Volumen, Biegewellenanregung, Abstrahlung, Absorption, etc.), letzteres geht von Gleichungen für diffuse Schallfelder aus und kommt letzten Endes zum gleichen Ergebnis wie eine statistische Energieanalyse erster Ordnung. Das Verfahren nach EN 12354 liefert die Eingangsdaten für den verwendeten Auralisationsalgorithmus.

Grundlagen der Auralisation

Es werden die Signalverarbeitung, die Binauraltechnik und Aspekte der Präsentation binauraler Signale beleuchtet. Wichtige Algorithmen der Signalverarbeitung sind hier z.B. die Interpolation von Terzspektren (Schalldämmmaß), um Frequenzspektren zur weiteren

Verarbeitung zu erhalten. Die Auralisation berücksichtigt die beidohrige (binaurale) Verarbeitung im menschlichen Gehör, daher werden kurz die Grundkonzepte der Binauraltechnik erläutert (Außenohrübertragungsfunktionen, Kunstköpfe, Frei- und Diffusfeldentzerrung). Abschließend wird auf Aspekte der Wiedergabe der auralisierten Signale eingegangen. Speziell bei der bauakustischen Situation sind die zu präsentierenden Signale sehr leise, besonders wenn eine relativ hohe Schalldämmung auralisiert werden soll. Daher sind Einflüsse der verwendeten Hardware zur Wiedergabe zu berücksichtigen.

Algorithmus zur Auralisation der Luftschalldämmung

Es wird der Algorithmus zur Berechnung der Schallsignale an den Ohren eines Hörers im Empfangsraum beschrieben, wobei die Schallquelle sich im Senderaum befindet. Ausgehend von den Standard-Schallpegeldifferenzen der Direkttrennwand sowie der flankierenden Wände und Daten zur Geometrie des Empfangsraumes werden Filter für die Schalltransmission sowie der Nachhallprozess im Empfangsraum modelliert. Die Situation im Empfangsraum wird folgendermaßen abgebildet: Ein Hörer befindet sich ungefähr in der Mitte des Raumes und schaut auf die Trennwand zum Senderaum. Die 5 abstrahlenden Wände werden als Punktquellen jeweils in deren Mitte modelliert. Aus den Standard-Schallpegeldifferenzen für die einzelnen 5 Flanken werden akustische Filter entworfen, welche die Schalltransmission nachbilden. Die Standard-Schallpegeldifferenzen liegen zunächst vor als Werte im Abstand einer drittel Oktav zwischen 50 Hz und 5000 HzĎurch eine Interpolation und Hinzufügen eines geeigneten Phasenganges wird ein zur weiteren Verarbeitung geeignetes Frequenzspektrum erzeugt. Die Richtung der Quellen und die Entfernung vom Hörer wird durch Außenohrübertragungsfunktionen und Zeitverzögerungen berücksichtigt. Die 5 Punktschallquellen strahlen also Schall ab, den der Hörer direkt empfängt. Das zusätzlich erzeugte Raumschallfeld (Reflexionen, Nachhall) wird berücksichtigt durch die Modellierung eines Nachhallvorgangs, welcher aus einer gemessenen Raumimpulsantwort abgeleitet wird. Die Raumimpulsantwort wird durch den Transmissionsgrad der gesamten Bausituation angeregt und bildet zusammen mit dem Direktschall aus den

5 Punktquellen eine Impulsantwort für die Übertragung zwischen Sende- und Empfangsraum. Ein beliebiges Mono-Schallsignal wird mit dieser Impulsantwort so verarbeitet, dass das Signal an den Ohren des Empfängers erzeugt wird und per Kopfhörer am PC abgehört werden kann. Die einzelnen Verarbeitungsschritte werden anhand von Formeln und Abbildungen beschrieben.

Trittschall

Die Modellierung der Trittschalldämmung ist ungleich schwieriger, da eine Rückwirkung zwischen Quelle (z.B. gehende Person) und Struktur (Deckenkonstruktion) berücksichtigt werden muss. Die Quelle wird modelliert als reale Kraftquelle bestehend aus idealer Kraftquelle und einer Innenimpedanz. Nach der Darstellung einer Methode zur Bestimmung der Quellparameter im statischen Fall wird ein verbesserter Ansatz zur Bestimmung der Parameter im dynamischen Fall beschrieben. Die reale wird charakterisiert durch zwei Messungen an unterschiedlichen Impedanzen. Aus den gemessenen Größen können die Parameter Leerlaufkraft und Innenimpedanz bestimmt werden. Anhand von Beispielmessungen wird die Methode verifiziert. Abschließend wird ein vereinfachter Ansatz zur Trittschallauralisation beschrieben, welcher die Rückwirkung der Struktur auf die Quelle vernachlässigt. Dieser Ansatz ist zulässig, wenn die Strukturimpedanz um Größenordnungen höher ist, als die Quellimpedanz. Ausgehend von üblichen Größen zur Beschreibung der Bauprodukte (Trittschallpegel) wird der Schall an den Ohren eines Hörers im Empfangsraum auralisiert. Die Modellierung des Epfangsraums entspricht dabei der Methode, wie sie auch bei der Luftschallauralisation verwendet wurde.

Verifikation

Dieses Kapitel beschreibt eine ausführliche Verifikation des Algorithmus zur Auralisation der Luftschalldämmung. Zunächst wird die korrekte Reproduktion der Eingangsdaten (Standard-Schallpegeldifferenz) überprüft. Dazu wird das auralisierte Signal sowie das

Senderaumsignal mittels eines Schallpegelanalysators gemessen. Aus den Schalldruckpegeln im Sende- und Empfangsraum und der Nachhallzeit des Empfangsraums wird dann die „auralisierte Standard-Schallpegeldifferenz" bestimmt. Es zeigt sich, dass nur geringfügige Abweichungen zwischen Eingangsdaten und gemessenen Daten auftreten. Zur weiteren Auralisation wurde die Schalldämmung einer realen Bausituation nach ISO 140-4 (an mindestens 5 Punkten in Sende- und Empfangsraum) gemessen und als Datensatz in die bauakustische Simulationssoftware eingegeben. Eine Auralisation an diesen Punkten und die Bestimmung des Schalldämmmaßes der Bausituation zeigt eine sehr gute Übereinstimmung mit der an der realen Bausituation gemessenen Schalldämmung. Des weiteren wurden Signale im Senderaum abgespielt, im Empfangsraum aufgezeichnet und parallel dazu auch auralisiert. Zusätzlich zur Auralisation wie in Kapitel 4 beschrieben wurde eine vereinfachte Auralisation verwendet, welche nur die Schalldämmung der Wand durch akustische Filter modelliert und das Schallfeld im Empfangsraum (Reflexionen, Nachhall) vernachlässigt. Ein Hörversuch mit den „realen" und auralisierten Signalen zeigte eine gute Reproduktion der Situation sowie eine Präferenz zur detaillierten Auralisation gegenüber einer vereinfachten Auralisation.

Anwendungen

Die praktische Relevanz des Verfahrens wird anhand zweier durchgeführter Anwendungen des Algorithmus zur Luftschall-Auralisation verdeutlicht. Die erste Anwendung befasst sich mit der Verständlichkeit von Sprache, welche durch schalldämmende Konstruktionen übertragen wird. Der Bezug zur Praxis liegt hier z.B. beim akustischen Komfort (Privatsphäre in Wohnungen) und der Wahrung der Vertraulichkeit in Büroräumen. Zur Ermittlung der Sprachverständlichkeit in Abhängigkeit vom bewerteten Schalldämmmaß wurden Hörversuche durchgeführt. Ein Verständlichkeitstest bestehend aus einfachen Sätzen diente als Senderaumsignal für die Auralisation. Die im Empfangsraum auralisierte Sprache wurde dann Versuchspersonen dargeboten, welche die verstandenen Wörter wiederholten. Es wird deutlich, dass aus dem Einzahlkennwert des Schalldämmmaßes keine Vorhersage der Sprachverständlichkeit getroffen werden kann. Basierend auf Berechnungen verschiedener

Sprachverständlichkeitsindizes wird ein Verfahren zur Angabe der Verständlichkeit anhand der frequenzabhängigen Schalldämmung vorgeschlagen. Die Durchführung der Hörversuche sowie die Grundlagen zur Ermittlung von Kennwerten zur Sprachverständlichkeit wird im Detail beschrieben.

Des weiteren wurde die Beeinflussung der mentalen Verarbeitung visueller Inhalte (Arbeitsgedächtnis) durch sprachlichen Hintergrundschall untersucht. Vor allem in Büroräumen spielt diese Situation eine große Rolle. In früheren Untersuchungen zeigte sich, dass die Störwirkung eines Hintergrundgeräusches auf die mentale Verarbeitung kaum vom semantischen Gehalt des Hintergrundgeräusches abgängig ist, d.h. dass z.B ein japanischer Sprecher im Hintergrund ebenso stört wie ein deutscher Sprecher. Anhand verschiedener auralisierter deutscher und japanischer Sprache wurde in Hörversuchen ein Zusammenhang zwischen Verständlichkeit, Störfaktor und semantischem Gehalt entdeckt, welcher jedoch quantitativ noch genauer zu untersuchen ist.

Zusammenfassung

Das Verfahren der bauakustischen Auralisation lässt sich in vielerlei Hinsicht einsetzen. Zum einen ermöglicht es dem bauakustischen Planer, dem Klienten die Wirkung einer Maßnahme zu verdeutlichen, zum anderen ist ein Einsatz in der Lehre sinnvoll. Studenten der Akustik, der Architektur und des Bauingenieurwesens können sich die Wirkung des Einsatzes bestimmter Baustoffe und bestimmter schalltechnischer Maßnahmen unmittelbar anhören. In der Forschung erlaubt es die Auralisation, umfangreiche Hörversuche durchzuführen, ohne den aufwendigen Weg der Beschaffung von Quellsignalen (Messung der Schalldämmung, Aufnahmen in Sende- und Empfangsraum, etc.) beschreiten zu müssen. Vor allem hinsichtlich der Ermittlung verbesserter Bewertungskurven für die Berechnung von Einzahlkennwerten aus frequenzabhängigen Daten mittels Hörversuchen ist die bauakustische Auralisation ein sinnvolles Werkzeug.

Appendix

Abbreviations – Terms – Symbols

Table 1: Abbreviations

Abbreviation	Description
ANSI	American national standards institute
CAC	Class of acoustic comfort
DIN	Deutsches Institut für Normung (German institute for standardisation)
EN	Europäische Norm (European standard)
IACC	Interaural cross correlation coefficient
MTF	Modulation transfer function
SII	Speech intelligibility index
STI	Speech transmission index
VDI	Verein Deutscher Ingenieure (German association of engineers)

Table 2: List of symbols

Symbol	Description	Unit
a	Equivalent absorption length	m
c_g	Group velocity	m/s
d_{ij}	Vibration reduction factor	
m	Mass	kg
$m(F)$	Modulation transfer function (MTF)	
n_i	Modal density in subsystem i (SEA)	
f	Frequency	1/s
k	Wave number	1/m
p	Sound pressure	N/m^2
v	Particle velocity	m/s
A	Equivalent absorption area	m^2
C	Spectrum adaptation term for indoor noise	

Symbol	Description	Unit
C_{tr}	Spectrum adaptation term for traffic noise	
D	Level difference between 2 rooms	dB
D_n	Normalised level difference	dB
D_{nT}	Standardised level difference	dB
$D_{v,ij}$	Vibration level difference	dB
E	Energy	N m
F	Force	N
F	Modulation frequency (Chapter 7)	N
K_{ij}	Vibration reduction index	
$L_{A,F,max}$	A weighted maximum level with time constant 'fast'	dB(A)
$L_n, (L_{n,w})$	(Weighted) normalised impact sound level	dB
$L'_n, (L'_{n,w})$	(Weighted) normalised impact sound level	dB
$L_{nT}, (L_{nT,w})$	(Weighted) standardised impact sound level	dB
L_R	Sound pressure level in receiving room	dB
$L'_{n,w,eq}$	Equivalent weighted normalised impact sound level	dB
L_S	Sound pressure level in source room	dB
ΔL_w	Impact sound level improvement	dB
$R, (R_w)$	(Weighted) sound reduction index	dB
$R', (R'_w)$	(Weighted) apparent sound reduction index	dB
R_A	Sound reduction index with the spectrum adaptation term C applied to the single number weighted sound reduction index R'_w	dB
$R_{A,tr}$	Sound reduction index with the spectrum adaptation term C_{tr} applied to the single number weighted sound reduction index R'_w	dB
S	Area of direct partition between 2 rooms	m^2
T	Reverberation time	s
T_s	Structure-borne reverberation time	s
W	Power	N m/s
Z	Impedance	kg/s
η_n	Internal loss factor of a subsystem (SEA)	m^2
η_{ij}	Coupling factor for power flow from subsystem i to j (SEA)	m^2
ρ	Air density	kg/m^3
σ	Radiation efficiency	
τ	Transmission coefficient	

Curriculum vitae

01.12.1970	Born in Aurich, Germany

Education

1977-1981	Primary school, Strackholt
1981-1983	'Orientierungsstufe' (Test for further education), Großefehn
1983-1990	Secondary school, Aurich
1990-1991	Civil service, Wittmund

Course of studies

1991-1997	Studies of electrical engineering at RWTH Aachen University
1997	Diploma thesis at the Institute of Technical Acoustics, RWTH Aachen University
1997-2005	Ph.D. at RWTH Aachen University

Employments

1997-2005	Scientific employee at the Institute of Technical Acoustics (ITA), RWTH Aachen University

Danksagung

An der Entstehung dieser Arbeit war eine Vielzahl von Personen direkt und indirekt beteiligt, denen ich im folgenden meinen Dank aussprechen möchte.

Für die Betreuung dieser Arbeit danke ich Herrn Prof. Dr. rer. nat. Michael Vorländer, ohne dessen Übersicht und Motivationsfähigkeit diese Arbeit nicht möglich gewesen wäre.

Für die Übernahme des Koreferats und die ungewöhnlich schnelle Durchsicht danke ich Herrn Prof. Dr.-Ing. Peter Vary.

Allen Mitarbeitern des Instituts für Technische Akustik danke ich für die Unterstützung und das freundschaftliche, kreative, wunderbare Arbeitsklima.

Besonders hervorheben möchte ich Michael Makarski, der mich mit seiner Freundschaft, Ideen, Motivation und Kaffee unterstützt hat.

Malte Kob sei gedankt für seine Freundschaft, Humor und Kaffee.

Die Werkstätten des Institutes lieferten mir auf wundersame Weise aus vagen Ideen und Zeichnungen innerhalb kürzester Zeit wertvolle Versuchsaufbauten.

Meine Diplomanden Jörg Holländer und Thomas Brandner sowie meine studentischen Hilfskräfte Jan Sokoll und Lisa Rotenberg lieferten mir wertvolle Arbeiten zur Sprachverständlichkeit und zur Trittschallauralisation.

Jürgen Graf opferte Abende und Nächte für die Messung von Schalldämmmaßen und Nachhallzeiten und die Aufzeichnung musikähnlicher Klänge in Büroräumen des ITA.

Sabine Schlittmeier von der Katholischen Universität Eichstätt-Ingolstadt danke ich für die schöne Zusammenarbeit zum Irrelevant Speech Effect.

Heinrich Metzen sei gedankt für die bauakustische Planungssoftware Bastian sowie für unbezahlbare Tips, Literatur und Diskussionen.

Alfred Schmitz, Werner Scholl und Heinrich Bietz danke ich für fruchtbare (nicht furchtbare) Diskussionen und Möglichkeiten zur Messung der Trittschalldämmung und Aufzeichnung von Schallsignalen an der PTB Braunschweig.
Verena Unmüßig danke ich unendlich für ihre Liebe, die mich auch durch den Panzer aus Stress und Nervosität hindurch erreicht hat.
Last but not least danke ich nochmals Malte Kob und Michael Makarski, dass sie mich vor dem Irrweg bewahrt haben, diese Arbeit mit Microsoft Word zu schreiben. Wer immer dieses liest, bevor er seine Arbeit schreibt, möge sich meine Worte einprägen. Hinterher ist es zu spät.

Bibliography

[AE03] E. R. E. Abou-Elleal, *Raumakustik - Interaktion visueller und auditiver Wahrnehmungen*, Ph. D. thesis, RWTH Aachen, 2003.

[ANS97] ANSI, *American national standard methods for calculation of the speech intelligibility index*, ANSI, New York, 1997.

[BFSS02] S. Blessing, H. M. Fischer, M. Schneider and M. Späh, *Besonderheiten bei der Berücksichtigung von Stoßstellen im Massivbau*, Fortschritte der Akustik - DAGA (Bochum, Germany), DEGA, 2002, pp. 527–528.

[Bra86] J. S. Bradley, *Predictors of speech intelligibility in rooms*, Journal of the Acoustical Society of America **80** (1986), no. 3, 837–845.

[Bra03] T. Brandner, *Characterisation of structure-borne sound excitation of floors (in German)*, Diplomarbeit, Institute of Technical Acoustics, RWTH Aachen University, February 2003.

[BSFS01] S. Blessing, M. Späh, H. M. Fischer and M. Schneider, *Verifizierung des Rechenverfahrens für die Luftschalldämmung nach EN 12354-1 für den Massivbau, Teil2: erreichbare Genauigkeit*, Fortschritte der Akustik - DAGA (Hamburg, Germany), DEGA, 2001, pp. 214–215.

[CP69] M. J. Crocker and A. J. Price, *Sound Transmission using Statistical Energy Analysis*, Journal of Sound and Vibration **9** (1969), 460–486.

[CP70] M. J. Crocker and A. J. Price, *Sound Transmission through Double Panels Using Statistical Energy Analysis*, Journal of the Acoustical Society of America **47** (1970), 683–693.

[Cra82] R. J. M. Craik, *The Prediction of Sound Transmission through Buildings using Statistical Energy Analysis*, Journal of Sound and Vibration **82** (1982), 505–516.

[CS00] R. J. M. Craik and R. S. Smith, *Sound Transmission through Double Leaf Lightweight Partitions Part I: Airborne Sound*, Applied Acoustics **61** (2000), 223–245.

[CSE91] R. J. M. Craik, J. A. Steel and D. I. Evans, *Statistical Energy Analysis of Structure-Borne Sound Transmission at Low Frequencies*, Journal of Sound and Vibration **144** (1991), no. 1, 95–107.

[Dat99] Datakustik (http://www.datakustik.de), *BASTIAN software for calculating sound insulation*, 1999.

[DIN89] DIN, *Sound insulation in buildings*, DIN, 1989.

[Elm80] A. Elmallawany, *Criticism of Statistical Energy Analysis for the Calculation of Sound Insulation: Part 2 - Double Partitions*, Applied Acoustics **13** (1980), 33–41.

[EN197] *EN 12354 part 1, Building acoustics - Estimation of acoustic performance of buildings from the performance of products*, 1997.

[FDS03] H. M. Fischer, A. Drechsler and J. Scheck, *Trittschall von Montagetreppen - Wege zu einer praxisgerechten Beurteilung der schalltechnischen Eigenschaften*, Fortschritte der Akustik - DAGA (Aachen, Germany), DEGA, 2003.

[Gen84] K. Genuit, *Ein Modell zur Beschreibung von Aussenohrübertragungseigenschaften*, Ph. D. thesis, RWTH Aachen University, Aachen, Germany, 1984.

[Ger79] E. Gerretsen, *Calculation of sound transmission between dwellings by partitions and flanking structures*, Applied Acoustics **12** (1979), 413–433.

[Ger86] E. Gerretsen, *Calculation of airborne and impact sound insulation between dwellings*, Applied Acoustics **19** (1986), 245–264.

[Ger90] E. Gerretsen, *Een Rekenmodel voor Contactgeluidisolatie bij verende Vloerafwerkingen*, Rapport no. 38, Stichting Bouw Research, 1990.

[Ger94] E. Gerretsen, *European developments in prediction models for building acoustics*, Acta Acustica **2** (1994), 205–214.

[GG76] B. M. Gibbs and C. L. S. Gilford, *The Use of Power Flow Methods for the Assessment of Sound Transmission in Building Structures*, Journal of Sound and Vibration **49** (1976), 267–286.

[Gus02] R. Guski, *Status, Tendenzen und Desiderate der Lärmwirkungsforschung zu Beginn des 21. Jahrhunderts*, Zeitschrift für Lärmbekämpfung **49** (2002), no. 6, 219–232.

[Hol02] J. Holländer, *Improvements on an algorithm for insulation of airborne sound insulation (in German)*, Institute of Technical Acoustics, RWTH Aachen University, 2002.

[Hol03] J. Holländer, *Modelling of Speech intelligibility in cases of sound transmission through sound-insulating constructions (in German)*, Diplomarbeit, Institute of Technical Acoustics, RWTH Aachen University, April 2003.

[HS73] T. Houtgast and H. J. M. Steeneken, *The Modulation Transfer Function in Room Acoustics as a Predictor of Speech Intelligibility*, Acustica **28** (1973), 66–73.

[HS84] T. Houtgast and H. J. M. Steeneken, *A multi-lingual evaluation of the Rasti-method for estimating speech intelligibility in auditoria*, Acustica **54** (1984), 185–199.

[HS85] T. Houtgast and H. J. M. Steeneken, *A review of the MTF concept in room acoustics and its use for estimating speech intelligibility in auditoria*, Journal of the Acoustical Society of America **77** (1985), no. 3, 1069–1077.

[ISO97] ISO, *DIN EN ISO 717 - Acoustics - Rating of sound insulation in buildings and of building elements*, ISO, 1997.

[ISO98] ISO, *EN ISO 140-4 - Acoustics - Measurement of sound insulation in buildings and of building elements - Part 4: Field measurement of airborne sound insulation between rooms*, 1998.

[ISO03] ISO, *prEN ISO 140-11 - Acoustics - Measurement of sound insulation in buildings and of building elements - Part 11: Laboratory measurements of the reduction of transmitted impact sound by floor coverings on lightweight reference floors*, November 2003.

[JHN04] A. C. Johansson, P. Hammer and E. Nilsson, *Prediction of Subjective Response from Objective Measurements Applied to Walking Sound*, Acta Acustica united with Acustica **90** (2004), no. 1, 161–170.

[JJVT04] J. Y. Jeon, J. H. Jeong, M. Vorländer and R. Thaden, *Evaluation of floor impact sound insulation in reinforced concrete buildings*, ACTA ACUSTICA united with ACUSTICA (2004).

[KH93] M. Klatte and J. Hellbrück, *Der Irrelevant Speech Effect: Wirkungen von Hintergrundschall auf das Arbeitsgedächtnis*, Zeitschrift für Lärmbekämpfung **40** (1993), 91–98.

[Kut73] H. Kuttruff, *Room Acoustics*, 1st ed., Elsevier, 1973.

[Lan97] J. Lang, *Ermittlung von Einzahlangaben für die Luft- und Trittschalldämmung und die Schallabsorption*, wksb **42** (1997), no. 40, 7–12.

[LM62] R. H. Lyon and G. Maidanik, *Power Flow between Linearly Coupled Oscillators*, Journal of the Acoustical Society of America **34** (1962), 623–639.

[LS02] T. Lentz and O. Schmitz, *Realisation of an adaptive cross-talk cancellation system for a moving listener*, 21st Audio Engineering Society Conference (St. Petersburg), AES, AES, 2002, p. 279.

[LSB98] K. Lehn, H. Strauss and J. Blauert, *Binaurale Auralisierung von Lärmschutzmaßnahmen in geschlossenen Räumen*, Fortschritte der Akustik - DAGA (Zürich), DEGA, 1998, pp. 542–543.

[Met92] H. A. Metzen, *Methoden zur Beurteilung des Luft- und Trittschallschutzes in den europäischen Ländern*, wksb (1992), no. 31, 1–16.

[Met97] H. A. Metzen, *Berechnungsmethoden für die Luft- und Trittschalldämmung in Gebäuden*, wksb **42** (1997), no. 40, 17–23.

[Met99] H. Metzen, *Accuracy of CEN-prediction models applied to German building situations*, Forum Acusticum (Berlin), DEGA / ASA, 1999.

[MM01] S. Müller and P. Massarani, *Transfer-Function Measurement with Sweeps*, Journal of the Audio Engineering Society **49** (2001), no. 6, 443.

[OW02] J. Ortscheid and H. Wende, *Lärmbelästigung in Deutschland*, Zeitschrift für Lärmbekämpfung **49** (2002), no. 2, 41–45.

[Pet02] E. Petzold, *Charakterisierung von Körperschallquellen im Zusammenhang mit der Anregung von leichten Treppen*, Fortschritte der Akustik - DAGA (Bochum, Germany), DEGA, 2002.

[Poh03] R. Pohlenz, *Bauordnungsrechtlich vorgeschriebener vs. erwarteten Schallschutz - zu recht enttäuscht?*, Fortschritte der Akustik - DAGA (Aachen, Germany), DEGA, 2003, pp. 130–131.

[Ras04] B. Rasmussen, *Sound insulation between dwellings - Classification schemes and building regulations in Europe*, InterNoise (Prague, Czech Republic), 2004.

[RR96] B. Rasmussen and J. H. Rindel, *Wohnungen für die Zukunft: Das Konzept des akustischen Komforts und welcher Schallschutz von den Bewohnern als zufriedenstellend beurteilt wird*, wksb **42** (1996), no. 38, 4–11.

[Sar04] E. Sarradj, *Energy-based Vibroacoustics: SEA and Beyond*, CFA / DAGA (Strasbourg, France), SFA / DEGA, 2004.

[SBF01] M. Späh, S. Blessing and H. M. Fischer, *Verifizierung des Rechenverfahrens für die Luftschalldämmung nach EN 12354-1 für den Massivbau, Teil1: Einfluß von Eingangsgrößen*, Fortschritte der Akustik - DAGA (Hamburg, Germany), DEGA, 2001, pp. 212–213.

[Sch93] A. Schmitz, *Naturgetreue Wiedergabe kopfbezogener Schallaufnahmen über zwei Lautsprecher mit Hilfe eines Übersprechkompensators*, Ph. D. thesis, RWTH Aachen, Shaker Verlag Aachen, 1993.

[Sch95] A. Schmitz, *Ein neues digitales Kunstkopfmeßsystem*, ACUSTICA **81** (1995), 416–420.

[SF03] A. Schmitz. and H. M. Fischer, *Sound Insulation Quality in Germany*, Fortschritte der Akustik - DAGA (Aachen, Germany), DEGA, 2003, p. 130.

[SKF04] M. Schneider, K. Kohler and H. M. Fischer, *Influence of flanking transmission on impact sound insulation in solid multi-dwellings*, CFA/DAGA, SFA/DEGA, 2004.

[SM99] W. Scholl and W. Maysenhölder, *Impact Sound Insulation of Timber Floors: Interaction between Source, Floor, Coverings and Load Bearing Floor*, Journal of Building Acoustics **6** (1999), no. 1, 43–61.

[Spa34] F. Spandöck, *Akustische Modellversuche*, Annalen der Physik V 20, 1934, p. 345.

[ST] S. J. Schlittmeier and R. Thaden, *Irrelevant background speech does disturb: The contribution of speech intelligibility and semantic content of low background speech to the Irrelevant Speech Effect.*, to be published.

[SWSK02] W. Scholl, W. Weise and S. Stange-Kölling, *Anpassung des Trittschall-Normhammerwerks an die mechanischen Eigenschaften gehender Personen*, Fortschritte der Akustik - DAGA, DEGA, 2002, pp. 565–566.

[SZ97] A. Schmitz and J. Zier, *Auralisation von Schalldämmaßen*, Fortschritte der Akustik - DAGA (Kiel), DEGA, 1997, pp. 95–96.

[TB03] R. Thaden and T. Brandner, *Characterisation of floor impact sources*, Proc. InterNoise (Jeju), 2003.

[Tha] R. Thaden, *Sound examples: http://www.akustik.rwth-aachen.de/Forschung/Projekte/tritt_aura (english)*.

[Tha01] R. Thaden, *Ein Modell zur Auralisation der Trittschalldämmung*, Fortschritte der Akustik - DAGA (Hamburg, Germany), DEGA, 2001.

[Tha04] R. Thaden, *Auralisation of impact sound insulation*, Proceedings of the ICA (Kyoto, Japan), 2004.

[Vor03] M. Vorländer, *Auralization in Noise Control*, inter-noise (Jeju, Korea), 2003.

[VT00a] M. Vorländer and R. Thaden, *Auralisation of airborne sound insulation in buildings*, Acustica united with Acta Acustica **86** (2000), no. 2, 76–89.

[VT00b] M. Vorländer and R. Thaden, *Hörbarmachung von Schalldämmung in Gebäuden*, Zeitschrift für Lärmbekämpfung **45** (2000), no. 1, 169–173.

[ZF99] E. Zwicker and H. Fastl, *Psychoacoustics, Facts and Models*, Springer, 1999.